13/v

Water as a Productive Environment

C. F. HICKLING

CROOM HELM LONDON

Croom Helm Ltd, 2-10 St John's Road London SW 11

ISBN: 0–85664–062–X Hardback
0–85664–190–1 Paperback

5

Printed in Great Britain by
REDWOOD BURN LIMITED
Trowbridge & Esher

CONTENTS

Clear and cool, clear and cool,
By laughing shallow, and dreaming pool,
Cool and clear, cool and clear,
By shining shingle, and foaming weir,
Undefiled for the undefiled . . .

Charles Kingsley, *The Water Babies.*

Chapter 1 INTRODUCTION

Water is one of the four elements of the Ancients, with Air, Fire and Earth. Of the four, Water had the character of movement, instability, turbulence. It rolled from change to change, and there are numberless classical allusions in the literature of the world. This book will be based upon this character of movement and will treat the subject from the ecological aspect, by following water from the hills and mountains on which it falls, down the path of gravity to the torrents and hill streams, thence to the rivers, to the lakes, to the artificial lakes and dams which men build to economise and use water, and so to the estuaries. Here, it mixes with the salt water from which it came, and is followed in the tidal coastal environment, on the continental shelf, in the overlying pelagic zone, and on to the final consummation of its long journey in the deep Ocean (Figure 1).

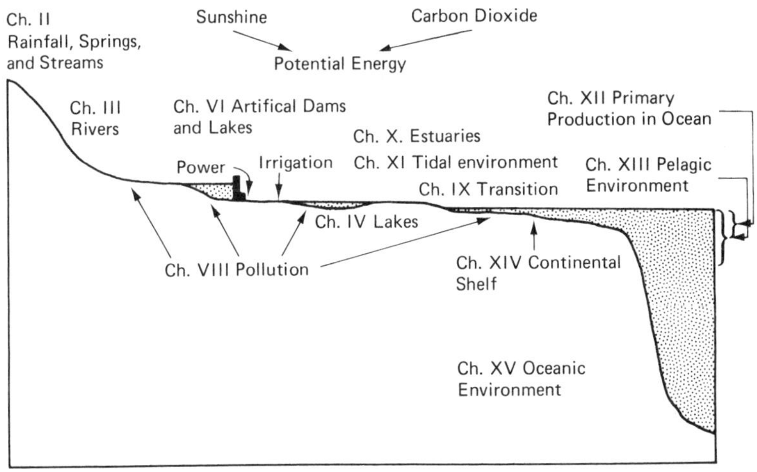

Figure 1 Plan of this book and arrangement of Chapters

The study of water is usually under two disciplines, namely, limnology, the study of inland waters which are usually, but not always, fresh; and oceanography. The plan of this book involves both disciplines, but seems a logical way of showing how water supports life in abundance in all the successive phases; what kinds of vegetation are responsible for this primary production, how much is produced, and how this primary production of material, which is also production of potential energy, fuels the activities of successive levels of animal life, leading to the fishes which provide man with food and recreation.

Chapter 8, placed between the land and sea chapters, considers pollution, a growing hazard and nuisance in the modern world, common to all phases of water, fresh and marine.

The Importance of Water

Water is one of the most important factors in the environment of man. He cannot escape it, for about 65 per cent of his own body weight is water, and he is continually losing it through skin, lungs, gut and kidneys. This water must be replaced without delay, or he will die. Man must dwell within reasonable reach of drinkable water, and wars have been fought for this priceless gift.

In advanced civilisations, water is something which comes from a tap and is extremely cheap; the cost is only a very few pence or cents per ton stored, processed, purified and delivered to the customer's premises. But the rate of consumption is rising rapidly, both with the increase of human populations and their changing customs. For example, modern methods of spray irrigation or artificial rain, resorted to especially in drought periods, call for large quantities of water most of which is evaporated or transpired so that very little returns to the rivers.

Fresh water is used over and over again in modern communities, purified for reuse. The water of the River Lea, on which London in part depends for water, is used several times over before being allowed to return to the sea. Modern technology has yet many devices for the salvage and reuse of water, as the experience of the astronauts has shown. None the less, a shortage is foreseen in the near future by the engineers of many advanced countries, and as more costly sources of water have to be drawn on, the price is bound to rise.

Water is not only an absolute necessity for the life of man and all the plants and animals on which he depends; it also provides much of his food, as fish and shellfish. Middens of bones and shells are associated with early settlements, so great that they have been mined for lime. The absence of water is death.

Before we pass on to the production of food, and energy, man is aided in two other ways by water. Firstly, water provides easy transport; for, with water bearing the weight, far heavier loads can be transported by rowing, dragging or poling, than could possibly be carried. Rivers are still main highways and transport routes. Secondly, as it descends from the hills to the sea, it provides man with a high proportion of the power he uses.

The Sun supplies our Energy

The sun provides us with waterpower: its heat has evaporated the water, and the winds it generates have carried it to the high ground from which it falls through our turbines. But its chief provision is potential energy, in the shape of carbon compounds which can be burnt, or more slowly oxidised, to give us back the energy fixed in its formation. This applies to all our fuels; for oil is the residue of vast algal growth in ancient freshwaters, and coal is the fossilised timber of the ancient swamp forests.

The Vital Reaction that fuels the World

The atmosphere of the earth contains a vast amount of the gas carbon dioxode. This has been estimated at 2,100 million million kilos, of which some 500 million million kilos is carbon. This quantity is being added to at the present time by the increased combustion of fuels, as well as from volcanoes.

Carbon dioxide is soluble in water, and from it the whole of the new organic matter in the world is made by a photochemical process in which the carbon dioxide is reduced to carbohydrate. This process is called photosynthesis, and through it the world stock of potential energy is built up. This is because the reduction of carbon dioxide to carbon, and thence to carbohydrate, needs a large input of energy, and this energy is obtained from sunlight. The living plant cells which perform photosynthesis contain chlorophyll, a group of pigments which together give plants their typical bright green colour. These pigments, alone or together with accessory pigments brown or red, which help to protect the cell from harmful excessive light, intercept the energy of the sunlight, thus carrying out the photoreduction of the carbon dioxide to carbohydrate, with the release of free oxygen, as in the formula below:

$$6\ CO_2 + 6\ H_2O \rightleftarrows 6\ C_6H_{12}O_6 + 6\ O_2\ .$$

Because a part of the energy of the sunlight has been used to accomplish this photoreduction, the product contains an equivalent amount of potential energy. If it is burnt, it gives back this energy as heat. But in the living plant, the organic matter can be made to give up its energy slowly and as needed. The energy so released can be used by the plant cells to reduce nitrogen-, phosphorus- and sulphur-containing compounds and incorporate these essential elements in building new protoplasm, or living matter. The process is the same in both water and land plants, for even in land plants the reaction goes on in water, namely in the sap in the cells.

The whole process has been summarised as follows: 1,300,000 calories of solar energy + 106 CO_2 + 90 H_2O + 16 $'NO_3$ + 1 $'''PO_4$ + traces of some mineral elements = 13,000 calories of potential energy, represented by 3,258 gm of protoplasm (106 C, 180 H, 46 O, 16N, 1 P, 815 gm of mineral ash) + 154 O_2 + 1,287,000 calories dispersed in calorific energy.

Clearly, this is not an efficient process. Only one per cent of the sun's energy, falling on the plant, is fixed as potential energy: yet this must drive the whole cycle of life in water. But a high proportion of the sunlight is reflected away at the water surface, and much of the light energy which enters is lost by absorption and reflection from small suspended articles. The above formula shows that most of the energy just warms up the water.

Photosynthesis can take place only in the presence of sunlight of the right strength and colour, and in the presence of chlorophyll. It requires a weak solution of carbon dioxide, a favourable ambient temperature, and traces of inorganic materials such as phosphates and simple nitrogen compounds.

It might be thought that the efficiency of photosynthesis might be greatly increased by a small increase in the efficiency of capture of light energy. But photosynthesis depends on other factors, and will proceed only as fast as the least abundant factor, just as the speed of a fleet is that of the slowest ship. Carbon dioxide is seldom in such short supply as to limit primary production, or photosynthesis. Hard and calcareous waters contain abundant lime and magnesium carbonates, which can take up carbon dioxide from the air and from the decay of organic matter, converting into bicarbonates from which the carbon dioxide can be readily split again for the use of plant life. Soft, acid waters are weakly productive, except in cases where fixed and rooted plants can extract nutrients from the soil, as happens in some Russian and American reservoirs and lakes.

The oceans contain abundant bicarbonates in solution.

Nutrient Salts

The large variations observed in the rates of primary production in water are mainly due to differences in the content of the nutrient salts, of which the most important are simple compounds of nitrogen and phosphorus, with sulphates (see the formula above), and silicates for the diatoms and traces of some essential metallic ions, such as potassium, magnesium, iron and calcium. These nutrients are present in freshwaters in quantities associated with the geology of the watershed in which they stand or flow. Alkaline or calcareous waters are the most productive. The alkaline bases are derived from chalk or limestone and from neutral or basic rocks such as schists, basic dykes, dolerites, lavas, metamorphic schists and gypsum. Waters flowing over gneiss and granites are acid and generally unproductive.[1,2] The nutrient salts such as nitrates and phosphates are derived from land drainage, or released into the water by the decay of dead plants and animals; this release takes place most favourably in neutral or alkaline waters. So well-known is this, that the water of fishponds can be made tenfold more productive of fish by neutralising with limestone and adding nitrates or ammonia, and especially phosphates.[3]

In the deeper waters of the ocean the nutrient salts are present in virtually unlimited supply below the surface layers accessible to sunlight. They have been accummulating there since the oceans existed. But this big reserve of nutrients can only be used when water movements bring them up into the zone of sunlight. The huge scale of the primary production which results reveals the vast resources of the ocean.

Light Penetration

The depth to which sunlight penetrates water varies with the altitude of the sun, the clearness of the water and the absence of shade. In the transparent waters of lakes and the open ocean, the light at the red and violet ends of the spectrum is absorbed in the first few metres. In fact, the red light is used up in warming the surface layers, as can be seen in the equation given on page 4. In progressing deeper, the yellow, green and blue light in the spectrum disappear in that order. In performing photosynthesis, waterplants use chiefly the green and blue light, of wavelengths between 4,200 and 5,600 Ångstrom units.

Like all other plants (and all animals for that matter) water plants respire, consuming oxygen and producing carbon dioxide day and night so long as the plant remains alive. Respiration is the reverse of the formula on page 3, and it oxidises a little of the carbon fixed during the hours of daylight to obtain the energy the plant needs to keep alive. It is much slower than photosynthesis, so that one hour's photosynthesis will supply many hours of respiration. In Arctic and Antarctic waters, plants have to fix enough carbon in the continuous hours of summer daylight to supply them with the energy to keep alive during the long and dark winter when no primary production is possible (p. 127).

But at a certain depth in the river, lake or ocean, the light becomes so weak that photosynthesis just balances respiration. Below this point, plants and animals behave alike, and both are consumers of oxygen and producers of carbon dioxide. The depth at which this happens depends on the colour and amount of suspended matter in the water, as well as on the angle of incidence of the sunlight. In turbid waters this depth may be between two and ten metres. It can happen that the growth of the plants is so intense that they shade out their own photosynthesis; while conversely, in the very clear waters of Lake Baikal, for instance, there is enough sunlight at seventy metres depth for at least some photosynthesis. So it can happen that a deep column of clear water thinly populated with plant life can produce as much fixed carbon as a shallow column of water turbid with plant growth.

The Kinds of Aquatic Plants

The green plants which carry out photosynthesis in water may be large in size and fixed to, or rooted in, the bottom or to some firm substrate. Such are the pondweeds in fresh water, and seaweeds in the sea. It has been found that their rate of primary production can be extremely high (p. 127).

But there are also small filamentous plants which grow closely attached to other plants, stones and rocks, dead branches and the like. They form a dense, slimy felt of algae or mosses, which are the most important primary producers in running streams. Their productivity is largely due to the rapid currents in which they grow; for these prevent the formation of a film of water, already stripped of its nutrient salts and carbon dioxide, around the plants, which would reduce their primary production. They may be very productive, the more so since the filamentous algae are themselves covered with attached diatoms. This was well shown in analyses of algal felts from pond bottoms. The felts had a high ash content, and of this 33 per cent was silica. Now the algae have no silica, so the big load of silica was derived from the dense epiflora of diatoms, which require silica, growing on the algae.

The whole complex of fixed algae and mosses, with their epiphytic plants and their dense infauna of small animals (p. 129), forms a characteristic habitat to which the German limnologists applied the word aufwuchs; no better term has been coined in English. It has been recommended by an eminent American limnologist that the word be taken over in its entirety; used without italics or inverted commas, this will be done in this book.

Both these kinds of plants grow, more or less, in two dimensions, and it is not difficult to measure their gross rate of production. The large attached plants can be cut repeatedly from a measured area, such as a square metre, and the cuttings weighed. The aufwuchs can be grown on artificial substrates such as glass slides, or tiles, and the rate of production measured by scraping off and weighing.

Productivity is usually defined in terms of weight of carbon produced in unit area in unit time or in the weight of oxygen evolved (see formula on p. 3). It is expressed as gm.C/m^2/day, or, as more often in American usage, as gm.C m^{-2} day^{-1}. In the case of lakes and the sea productivity is usually that taking place below one square metre of the surface. But often it is expressed as production per cubic metre.

But the third kind of green water-plant which carries out photosynthesis is far and away the most important. This is the phytoplankton, the vast assemblage of small to very minute unicellular plants which live suspended in the water, whether in slow rivers, in lakes or dams, or in the sea. As the ocean is incomparably larger in area than the freshwaters, almost the whole of the aquatic production of matter and energy is due to the phytoplankton, especially the marine phytoplankton.

As these minute plants are distributed in three dimensions, it is less easy to determine their productivity, very important though this is. The methods evolved have been applied to all phytoplankton, freshwater and marine. But as the last are by far the most important, the methods will be described in Chapter 12, which deals with the life of the upper waters of the oceans.

The Distribution of Water

Of the vast stock of water in the world, some 97 per cent is in the seas and oceans. Of the three per cent which is freshwater, about two per cent is locked away in the polar ice caps, leaving only about one per cent as free freshwater. All this freshwater is in transit back to the ocean from which it came. The freshwater in the ice caps is in very slow movement; great chunks fall off the margins as icebergs, and these slowly melt as they drift into warmer waters. There have been half-serious suggestions that these floating mountains of pure fresh water might some day be drawn upon to supplement dwindling supplies from other sources.

But the one per cent available as freshwater is in comparatively rapid movement, collecting into streams and rivers, and always trending downhill back to the ocean. Man can use it for only a short while as it passes him by.

World Winds and Ocean Currents

There would be no freshwater, but for a global process of distillation, in which seawater is evaporated and then condensed on the land. This process depends on the world system of winds and the ocean currents which they propel. Enormous forces are needed to move, often at considerable speeds, the billions of tons of water which comprise the ocean currents, and all this energy comes from the sun.

The cycle begins with the differing amounts of heat from the sun which reach the earth at the equator and at the poles. The heated air over the tropics tends to rise, while the cold air over the poles tends to sink; thus there is a very complicated net surface flow of air from the polar regions to the tropics. A second important effect is due to the difference in the capacity for heat of the different parts of the earth's surface, and especially the high capacity for heat of the oceans, which warm up only slightly when exposed to the heat which causes the land surface to warm up considerably. Thus the land becomes warmer than the sea in the summer, while the reverse prevails in winter, and this has a strong effect on the wind systems.

A generalised diagram of the wind systems, and of the ocean currents as they generate, is shown in Figure 2.[4]

There is a permanent belt of low pressure around the equatorial regions, with a mean position just north of the equator, between 5° and 10° north latitude. This belt moves north and south seasonally following the sun. The warm rising air is saturated with water, and tends to be pushed up by colder air from the subtropics, the so-called intertropical front; this gives heavy rainfall as the front passes across the equator and back.

Polewards of this low pressure belt are circular gyres around areas of high pressure. The winds blow in a clockwise direction around the northern gyre and in an anticlockwise direction around the southern gyre. On the equatorial side of each gyre, the winds blow from an easterly direction and into the low pressure belt, and are in fact the well-known trade winds, northeast and southeast respectively. They propel the surface water to the west as the north and south equatorial drifts or currents. These currents impinge on the eastern coasts of the continents, and generate return or countercurrents. An important one is the equatorial countercurrent, which flows eastward roughly along the equator and between the north and south equatorial drifts. Where the margins of these oppositely-flowing currents are in contact, powerful turbulence and eddies are set up, which have the very important result of bringing to the surface waters significant amounts of the deep and nutrient-rich waters from below the zone to which sunlight penetrates (the photic zone).

The water of the north and south equatorial drifts also escapes to the north and south along the continental coasts. The most important are the Gulf Stream and its Pacific counterpart, the Kuro Shio. These stream to the northeast as warm and fast currents, which are helped on their way by the westerly winds on the northerly edge of the northern gyres. They carry heat from the tropics to the temperate and even arctic zones. As they die out, their water is carried along by the winds as the North Atlantic Drift and the North Pacific Drift. Cold counter-

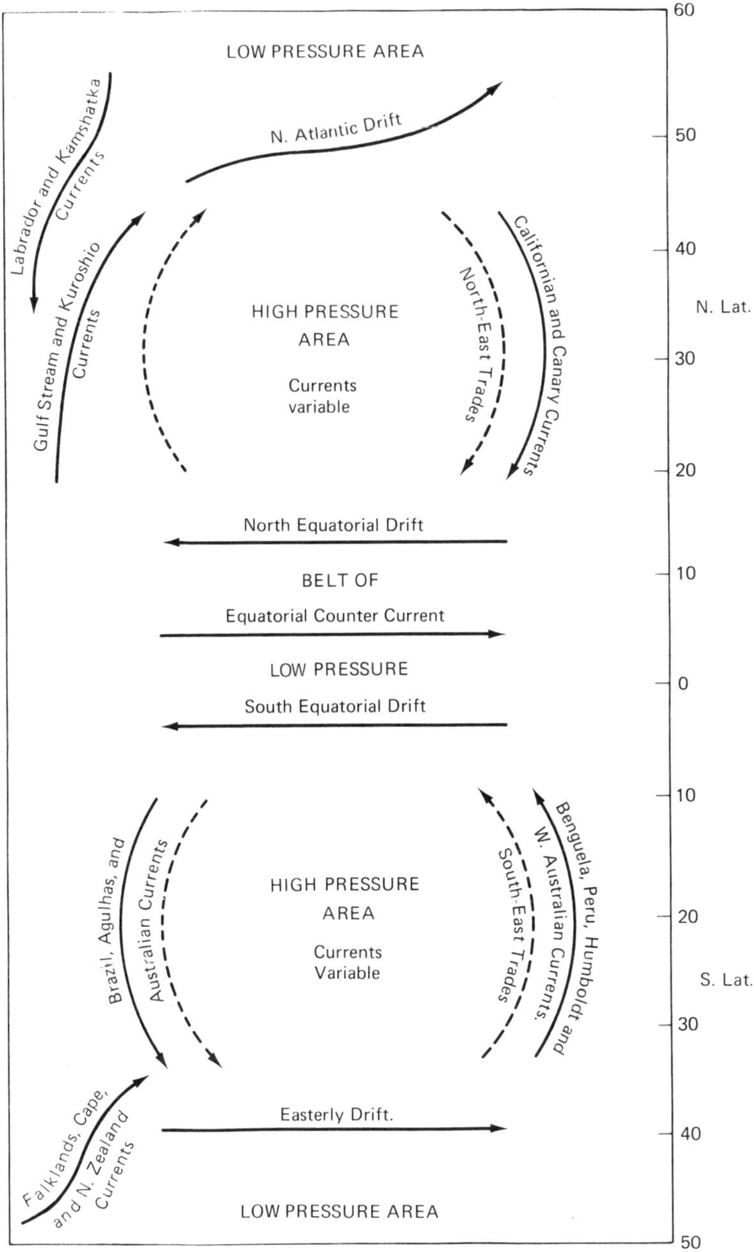

Figure 2 Diagram of winds and ocean currents[4]

currents, the Labrador and Kamschatka currents, flow from the Arctic and carry icebergs south into warm waters. Again, important areas of ocean fertility occur where these currents pass each other.

On the southern side of the southern gyre there is another belt of westerly winds, the famous 'roaring forties,' which drive a strong easterly drift to encircle the globe along the 40° and 50° south latitude. Branches of this drift are deflected north along the west coasts of South America (the Humboldt, Peruvian currents) of Africa (the Benguela current) and Australia (the West Australian current). These north-moving currents are helped along by the winds blowing up the eastern side of the gyre. On the western side, south-moving currents, partly generated by the equatorial drifts (the Brazil, Agulhas and Australian currents) are driven also by the south-moving winds of the southern gyre. As in the northern hemisphere, there are cold countercurrents (Falkland Islands and New Zealand currents). Within the gyres themselves, both winds and currents are variable and light. Because these gyres are regions of no mixing of water, their stores of nutrient salts are soon used up, and they are areas of low production.

These great ocean currents are not more than a few hundred metres deep, for the winds act on the surface, and friction with the underlying layers slows the current and also changes its direction. It has been found that countercurrents exist beneath the surface currents. In deep water, there are also currents, due to differences in the density of the water through variations in temperature and salinity. Even in the very deep ocean, ripples have been photographed in the bottom sand and mud revealing the presence of appreciable currents there.

Rainfall

The heavy rain caused by the forcing upwards by high ground of moisture laden winds is called orographic rain. In such cases, there is often a region of poor rainfall on the lee side, where there may be less than ten inches of rain. In the wet tropics, too, there is often convective rain; in the afternoon, the heated air rises and becomes cooled and heavy local rain falls. This may happen almost every afternoon, and naturally helps to keep the land well watered. In Britain, and western Europe generally, much of the rain is orographic; but there are also the cyclones or depressions which are always passing from west to east. These arise at the interface between the warm westerlies blowing across the top of the gyre and the cold easterlies in polar latitudes. Each cyclone has frontal systems in which the warm and saturated air is forced upwards by wedges of cold air and deposits its moisture as rain. As these depressions are travelling eastwards at all times of the year, rainfall in western Europe is plentiful and well-distributed. But in all continents, away from the sea, rainfall becomes small. In 1972 there was a failure of rainfall in the grain belts of India, Siberia and China, leading to a sharp rise in world grain prices as these countries had to enter the markets to buy. Variations in the ocean currents, due to a slight change of wind, can have catastrophic effects. In 1972, the Equatorial countercurrent flowed further south than usual; the desert coasts of Peru and Chile were deluged with rain which caused terrible damage, and the greatest fishery in the world, for the Peruvian Anchoveta, was a failure, providing only one million instead of the normal ten million tons. As these fish are used to make high grade fishmeal, this failure resulted in a rise in the world price of eggs, pork, veal, milk and poultry.

The Monsoons

In the Indian Ocean, the northern hemisphere consists of land; this causes a different wind and current regime from those of the Atlantic and Pacific Oceans. The southern gyre is present; but the northern gyre is replaced by the monsoon winds and their associated currents.

In the northern summer, the continental landmasses of Asia become hot, the heated air ascends and is replaced by cool and saturated air flowing in from the southwest over the Indian Ocean; the difference in pressure is often great enough to generate a gale force in these southwest winds, the so-called monsoon. As this wind crosses the coasts and is forced upwards by hills and then mountains, it is cooled and releases large quantities of rain and snow. There is one weather station in Assam which has an average annual rainfall of 460 inches, all of which falls during the monsoon. These massive rains are the source of many of the greatest rivers in the world. From the central Asian mountain mass, on to which the southwest monsoon wind deposits its rain and snow, flow the Indus, Ganges, Bramaputra, Chindwin, Irrawaddy and Mekong to the southward and the Indian Ocean; the Si Kiang, Yang tse Kiang, Yellow river and Amur drain to the Pacific, and the Lena, Ob and Yenisei to the Arctic. The Oxus and Jaxartes drain to the inland Aral Sea.

But in the winter months, the landmass becomes colder than the warm sea, so the air currents and the sea currents in the northern Indian Ocean reverse to give the cool, dry and pleasant northeast monsoon, which in fact is the northeast trade wind. Weather of a monsoon-like kind also occurs elsewhere. For example, in tropical West Africa the summer months bring windy rain from the southwest, off the ocean, often very heavy rain with thunderstorms. In the winter months, the winds are light from the north, often dust-laden from the Sahara; this wind is called the 'Harmattan.'

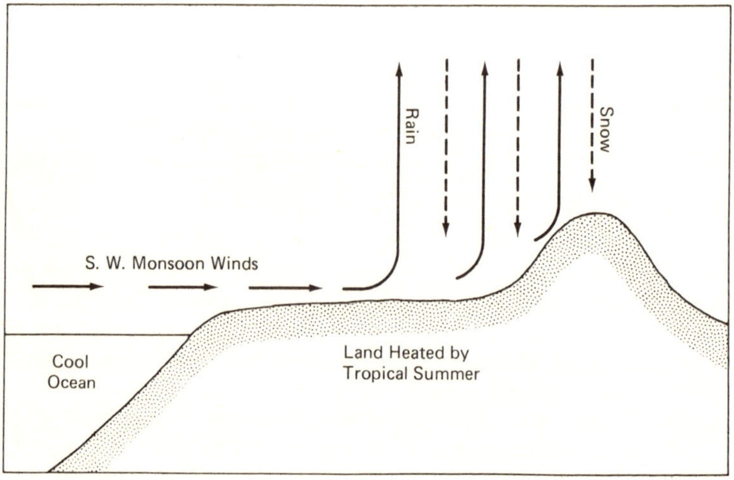

Figure 3 Diagram to show mechanism of monsoon winds and rain

It has been pointed out that the monsoon winds, with their seasonal reversal of direction, is the same phenomenon as the familiar land and sea breezes, on an annual and not a daily scale.

NOTES

1. Depasse, P. 1956. 'Monographie piscicole de la Province orientale.' *Bull. Agric. Congo Belge*, XLVII (4) 1956.
2. Braun, R. 1952. 'Limnologische Untersuchungen an einigen Seen im Amazonasgebeit.' *Schweiz. Z. Hydrol*, XIV 1952.
3. Hickling, C.F. 1971. *Fish Culture*. London, Faber & Faber, 1971.
4. Fowler, G.H. and Allen, E.J. 1928. *Science of the Sea*. Oxford, Clarendon Press.

Chapter 2 RAINFALL, SPRINGS AND STREAMS

Rainfall and melted snowfall flies away, runs away or soaks away. If it flies away as water vapour, it may be blown onwards to fall again as rain further downwind; in evaporating, it may have important cooling effects and it may also leave important concentrations of solutes.

Subsurface Water

When rainfall soaks away into porous soils, it will seep down below the level of the plant root systems until it reaches an impermeable layer of rock, where it accumulates to form an underground reservoir, the upper limit of which is called the 'water table.' Some water tables are very ancient. For example, Arabia and the Sahara were once well-watered, and over the millennia water accumulated in natural underwater reservoirs which can now be tapped by wells. The same is true of some of the Australian deserts. But it is well understood that this 'fossil' water cannot be replaced once it is drawn, unless there is a climatic change.

In dry countries, the rare rainfall which sweeps down the sandy river beds may sink down into the sand until it reaches the rocky river bed. There is is held, and may slowly seep its way downstream. Where a reef of rocks runs across the river bed, the water may be held to form a natural subsurface reservoir. The situations of these are of course well-known to the local inhabitants, who dig down to the water-saturated sand. These reefs can be imitated artificially by putting clay fillings in the stream bed. More than a half of the resulting mixture of sand and water may be water, and it will be protected from evaporation by the overlying sand.

Springs

Where the water sinks very deep, it may come under pressure, and will then rise to the surface with force when tapped by a pipe; or the water may be heated and emerge as a hot spring. Depending on their history since they sank underground and the nature of the rocks with which they have been in contact, they may be mineral springs. But in most cases the water emerges as freshwater, filtered and even oxygenated during its passage underground. Springs occur, of course, where the impermeable stratum which holds the water emerges at the surface of the ground, and they may be of three kinds. In rheocrene springs, the water flows away at once; in limnocrene springs, the water wells up into a pool, which overflows to run down the valley; and finally, the helocrene spring rises in marshy ground and seeps away; in this kind of spring the water may be acid and devoid of oxygen. Water from such a source will not support life until it has had time to dissolve air from the atmosphere. Rheocrene and limnocrene springs may also be rather deficient in oxygen, to the extent of only fifty to eighty per cent of saturation when they first emerge, but they are usually clear and have a constant temperature. Such springs can be used to supply comparatively warm water in very cold weather; fish such as *Tilapia* can be overwintered in northern Taiwan in ponds watered from springs even when the ambient temperature is below the lethal point for this tropical species. After a short course above ground, the water from these springs comes to have the same composition and oxygen content as water in streams which have run over the land. But spring water is most sought after for irrigating trout farms, if it is pure and at the right temperature.

A number of springs in Florida were examined for their productivity of plant material.[1] All were large artesian springs having a constant temperature and chemical composition, and thus were especially favourable for investigating the effect of light as the controlling factor in plant production. Three of the springs emerged with their water denuded of oxygen, no doubt having flowed through soils rich in organic matter; these springs were poorly productive, since the plants themselves could not live in anaerobic conditions, though they could soon produce some oxygen by photosynthesis. Their rate of production was measured at 0.5, 0.9 and 1.5 gm $O_2/m^2/day$. In a small shaded anaerobic spring on a rainy day, production was only 0.7 gm $O_2/m^2/day$. But at the other extreme, production was 64 gm $O_2/m^2/day$ on a sunny day in an aerobic spring well-grown with algae, themselves well covered with a dense aufwuchs of diatoms.

Subterranean Aquatic Life

Even while the water is still underground, it is not devoid of life, for it is a habitat free from predators, reasonably well oxygenated, and with enough organic matter to nourish a sparse fauna which can live in the capillary spaces within the gravel and sand through which the water percolates, as well as in the larger underground spaces of caves and streams. This infauna knows no seasonal climatic changes, and lives in total darkness. So the higher members of this group of animals, chiefly crustaceans, are blind and colourless. Best known is perhaps the fine pure white amphipod *Niphargus*, which very occasionally turns up in tapwater. This underground fauna includes some primitive forms, especially the crustaceans *Bathynella* and *Thermosbaena*. Neither of these two crustaceans fits into the standard classification. *Thermosbaena* lives in the hot springs which supply a set of public baths in Tunis. Some serpulid worms live in these situations, and there are also a few blind cave newts and fish.

Streams

If the rainwater runs away it forms surface runnels which coalesce into streams and hill torrents, and these into rivers. Plant and forest cover are of very great value in absorbing and releasing rainfall. Where there is no such cover, rains cause destructive floods which carry all before them. The coasts of Peru and northern Chile are deserts without vegetable cover; and every few years the ocean current system may change so that terrific rainstorms occur: as there is nothing to check the water, great devastation ensues. Destruction of forest cover is known to lead to desert conditions and loss of rainfall. In the Lake Baikal catchment, too extensive timber felling has led to the drying up, or intermittent flow, of many streams, affluents of the Lake; and, in the early days of the settlement of the watershed of the Great Lakes of North America, the same effect was observed (Chapter 5). Records from rubber estates in Malaysia show that, even locally, rainfall can be adversely affected by jungle felling and clearing, so that droughts become more frequent. This sets a dilemma for the undeveloped countries; they need more land for their growing populations, and yet, if they clear the forests, they risk losing some of the rainfall on which all else depends. Swamps and moorlands are also very valuable conservers of water. Swamps are full of vegetation in all stages of decay, the whole forming a sponge which can absorb great quantities of water and release it as a steady stream. Water can also be lost, however, from swamps; the best known case being the Sudd of the southern Sudan, an inland delta full of vegetation; an important part of the inflow of the White Nile is lost by evaporation in these swamps, so much so that plans for the future regulation of the Nile include a canal, the Jonglei Cut, through these swamps along which the White Nile can cross the swamp with the minimum loss. Peaty soils, spread over impervious rock, can hold much rainfall and release it slowly so as to keep moorland streams flowing the year round.

Streams which rise and flow through conifer forests may absorb toxic substances from the trees and litter: streams which issue from swamps may be very deficient in oxygen, and, like water from moorlands, may be deepy coloured brown or yellow with dissolved or colloidal organic matter. But this is usually soon coagulated by bases in the water, and then forms a floc which contributes to the detritus of the stream bed.

Composition and Productivity of the Water

The geological nature of the catchment much affects a stream's composition and fertility. Streams rising in stony or rocky country, devoid of much vegetable cover, may have a wide diurnal variation in temperature. They may also have a higher mineral content than other waters, since the heat of the sun will make the rocks more friable and so expose them to leaching by running water. A stream will also dissolve materials from the soil and rocks over which it flows, and these will affect its reaction and fertility. Studies on the headwaters of streams in the former Belgian Congo showed that streams flowing in areas of granite and gneiss had slightly acid waters, poorly buffered, with a very low capacity for supporting aquatic life; but streams flowing through areas of neutral or basic rocks were much richer in aquatic life.[2] In the basin of the Amazon, lakes in tertiary soils were uniformly poor in dissolved salts, while many lakes in the carboniferous were rich in electrolytes; the fauna of the bottom muds and the plankton of these lakes reflect these differences very clearly.[3]

Three streams in Perthshire, Scotland, which flow over and through different types of country, show notable differences in productivity, as may be seen in the Table below.[4]

Stream	Nature of Catchment	Conductivity μ mhos/cm^3	Productivity Dry wt of biomass gm/m^2
Almond	Quartzite and mica schist	46	0.65
Shelligan	Sandstone and quartzites	72	1.39
Fender	Limestone and black schists	150	2.04

The poor conductivity of the water in the Almond flowing over quartz and schists shows a low content of dissolved salts, and is reflected in a low standing biomass. The Fender shows the reverse of this, and the Shelligan is intermediate.

through open pasture land and open woodland of sycamore and elm trees, and also thorn bushes. The average width was 11.4 feet (3.5 m.), so that the total area of stream investigated was about 1.65 acres (0.67 ha.). The stream consisted of alternate riffles and deeper pools, the riffles having a total length of 2,298 feet (703 m.), and the pools of 4,010 feet (1230 m.).[9]

In production studies made in this stream, the freshwater 'shrimp,' *Gammarus fasciatus*, was the dominant organism. The wet weight of bottom-living animals varied little, on average, over the four years of the investigation, and amounted to about 5.047 gm/foot^2 wet weight, or 0.942 gm dry weight. This is about the equivalent of 10.07 gm/m^2. Reckoned on a per acre basis, production was about 486 lb. (432 kg./ ha.), which compares with the 321 lb./acre (286 kg./ha.) of the shallower parts of the fertile Lake Mendota, also in North America.

The rate of production of rainbow trout in this Virginian stream averaged 30.6 lb./acre/annum (27 kg./ha./annum) with little variation during the four years.

The Downstream Drift of Food Organisms

In the streams, an equally important source of food for the animals is the drift of organisms downstream from the beds of aufwuchs and from the silt communities in the stream bed. This drift is a very important source of food for fish and other animals. The blackfly larvae hold on to a firm substrate in the rushing water with a ring of hairs at their hinder end, and with the fore end catch drifting food particles. This downward drift of organisms is not wholly accidental, because it has a well-marked diurnal periodicity.[10] Of course, animals are always liable to be forced out of their niches by sheer population pressure, as well as by heavy spates after rainfall, though they will adopt a variety of devices to cling to their substrate. The bristles of the blackflies have been mentioned; other stream creatures avoid the force of the current by having flattened bodies which cling closely to the substrate, some so closely that they may even act as suckers. The familiar caddis fly larvae which make themselves cases of sand or plant debris have flat cases, with wing-like extensions, to present a low profile in fast water. Other and smaller animals can live and creep along the bottom in the very thin layer of water where friction reduces the speed of the current to almost nothing.

More insects are present in the downward drift at night than by day, and this may be the result of a greater activity then. The larvae of Trichoptera, for instance, move about and feed actively at night, and may thus be at a greater risk of being carried off then.

But it seems that a downward drift is a part of the life history of many stream organisms. By planting their young well upstream, organisms can secure for them the best conditions for feeding and growth. It is a curious analogy with fish which seem to ensure that their young are carried, or drift, into shallow water. Almost invariably, the small stages are found in shallow water, the larger stages in progressively deeper water. In the case of stream-dwelling organisms, this requires an upstream movement of the adults for breeding. In flying insects, the adult imagines fly to the headwaters of streams to lay their eggs; for example, the Trichoptera imagines emerge chiefly at night and fly upstream to lay their eggs. The youngest instars of the caddis flies are found most plentifully in the headwaters of a stream, and the older instars in the lower reaches.[11]

Upstream and Downstream Migrations

These upstream and downstream migrations have been studied in the English Lake District[12] and in Sweden.[13] Simple traps are placed in the stream, facing both ways, to catch creatures ascending and descending the stream. In the Lake District, the greatest number of migrants was taken at night; the numbers migrating upstream varied from 366 per 24 hours in November to 1690 in April. They were chiefly the nymphs of Plecoptera and Ephemeroptera, small larvae of diptera, and small individuals of the freshwater 'shrimp,' *Gammarus pulex*. The migrants mostly kept close to the banks and to the stream bed, and they took advantage of eddies in the lee of stones and in gravel.

But it was found that fewer individuals migrated upstream than downstream; upstream migration compensated for about one third of the downward drift in the winter months, and for only one tenth in the summer, when migrations were most active. However, a part of the upstream movement may be made by flying adult insects; and the upstream migrants consist of well-grown adults, of equal weight to many small young stages. This is certainly so with fish. In a stream in a temperate climate, the movements of fish were observed by the catches of traps, and it was seen that more fish moved downstream than upstream; but the upstream migrants were mature and full-grown fish, whereas the downstream migrants were juveniles. The average daily weight of fish caught on the downstream migration was 472 gm in 17 fish, at an average weight per fish of 28 gm. But the upstream migrants were 7 fish weighing 1081 gm, at an average weight of 155 gm.[14]

The upstream migration of stream animals of all kinds serves to recolonise areas denuded of life, as in the case of the South African stream which had its fauna poisoned by cattle dip. A study of fish population movements in small rivers has shown that, while most fish tend to remain within a radius of a few hundred metres, a proportion is mobile, and serves speedily to recolonise waterways which have lost their fish.[15]

But upstream migrants may play an important part in the chemistry of a stream. Fish may help to maintain useful reserves of phosphorus, for example, in the headwaters of streams, not only by their metabolism, but by the decay of their bodies when they die. The natural breeding streams and lakes of the Pacific salmon are unproductive and sterile; but the products of decomposition of the bodies of the exhausted salmon support a rate of production high enough to secure a fair survival of the young salmon when they hatch out of the redds.[16] The productivity of the stream and its rate of respiration may be actually higher near the headwaters than further downstream. In two streams in Carolina, for instance, the gross photosynthesis varied from 0.21 to nearly 9 gm $O_2 m^{-2} day^{-1}$, and community respiration from 0.4 to 13 gm $O_2 m^{-2} day^{-1}$.

Drifting Organisms as Fish Food

The downstream drift of organisms is an important source of food for fish. In a stream in northern Sweden, this drift amounted to as much as 37 to 288 gm/m^2/24 hours. In tropical streams it is equally important. In the headwaters of the Congo River, the downstream drift of insects, both aquatic and terrestrial, is the most important food for many species of *Barbus*. In Java, the downstream drift of insects from the sewage-enriched bottom fauna of streams is used by fish farmers to secure very rich feeding conditions for their fish. These streams, as they pass through towns, are used as public sewers and flood drains; consequently, among rotting masses of leaves and grasses such useful items of fish food as oligochaete worms and chironomid larvae were found to the extent of 329 worms and insect larvae in one square decimetre; equivalent to the very high figure of 70 gm/m^2.[17] From such rich populations a bountiful downstream drift occurs; so the fish culturists place small carp in cages staked to the stream bed, and find they grow fast. The fish can be very densely crowded in the cages because the fast stream flow secures enough oxygen; and if necessary (and profitable) the natural food can be supplemented with boiled rice and the like. The carp stocked are about 5 to 8 cm. in length, and as many as 200 to 400 fish may be stocked in a cage measuring 1.5 to 3 metres long, 90 to 150 cm. wide, and 60 to 70 cm. high. The fish attain a weight of 3.5 kg. and have the shape of fish fed on the optimum amounts of food. Net production can be as much as 50 to 75 kg. of fish per cage in two to three months.

Below a modern sewage works in the London area the bottom mud was found to contain 300,000 specimens of chironomid larvae, and

200,000 of the water louse *Asellus*, per m^2, besides abundant tubificid worms and free-living nematode worms.[18] Such a large community must send a rich drift downstream. Tubifex worms occur especially where there is some pollution of the water.[19] They are an important fish food, and in fact are sold by aquarists for that purpose. In streams, they are especially important as food for bream and rudd, and their importance in this respect may have been overlooked because they are so rapidly digested. But unluckily, these worms are also the intermediate hosts of the tape worm *Caryophyllaceus*, and the widespread occurrence of this parasite among coarse fish suggests that tubifex worms are food for most of them.[21]

Salmon and Trout

Young trout and trout fry find the drift of food organisms of great importance. The small fish can pick from the drift the size and quality of the food they need, keeping station in the stream as the food drifts down to them. Larger trout have the same habit, taking the emerging adults of benthic insects, as well as terrestrial invertebrates such as insects washed into the stream and worms washed in by bank erosion. But they also take nondrifting invertebrates such as caddis fly larvae; the fragments of plant debris and the small sand particles often found in the stomachs of trout are from the cases built by these larvae. Though the drift of invertebrates becomes progressively less important as the trout grows larger, the habit of watching for possible food morsels persists throughout life, and profits the dry fly fisherman.

Small salmon and trout are the most important fish found in Scottish hill streams, though eels are also present. They find their food among the invertebrates inhabiting the stream, such as stoneflies, mayflies, chironomid larvae and the adults and larvae of the beetle *Esolus*. These invertebrates find their food in and among the detritus in the pools, derived from disintegrating leaves and other organic remains, and also among the aufwuchs growing on the submerged stones in the riffles. It was found that the larger stones produced the more numerous food organisms, and that the algal felts growing on the undersides of stones contained more food organisms than in the beds of water mosses growing on the tops of the stones.

The standing stock of the bottom fauna of these small trout streams, Almond and Shelligan, was about 2.88 gm/m^2, dry weight, and this served to nourish a salmon-parr population of 6.5 to 11.1 gm/m^2, wet weight. The total trout populations in the same streams was 7.7 to 12.3 gm/m^2, wet weight, and trout and salmon parr together were being produced at a rate of 14 to 24 gm/m^2, wet weight. But in this total the yield of salmon parr was low.[21]

This low yield of salmon parr is due to the successful competition of the trout, which hatch out earlier in the season and are thus able to get a head start in establishing themselves in the stream. This was proved by an important experiment in which advanced salmon eggs were placed in the Fender Burn so that they hatched out before the trout eggs. The result was that the salmon parr now grew faster than the trout and had a much higher rate of survival than when the trout were allowed to dominate the stream. So well did the parr grow that by the end of their first year they were already beginning to take on a silvery sheen and would probably become smolts in their second year of life. As the salmon is commercially more valuable than the trout, a way seems in sight for increasing the production of salmon smolts in streams.[21]

Hard and Soft Water Streams

Where the streams have descended to more level ground, several other species of fish may be present, and there is a contrast between hard water streams, which flow through chalk country and contain dissolved lime, and soft water streams, slightly acid and with little or no lime. The hard water streams are more productive than the soft water streams; it has already been said that the composition of the water is affected by the nature of the soil over which the streams run, and that this also affects their fertility (pp. 4 and 5).

The rates of production, growth and survival of six species of small fish were measured in one soft water and five hard water streams in southern England. The principal fish in the streams were trout and bullhead or sculpin (*Cottus sp*), the former dominating in the soft water streams, the latter in the hard water streams. Trout made 87 per cent of the production of the soft water streams, the rest being chiefly minnow; in the hard water streams, the bullhead made 43 to 83 per cent of production, and trout only 7 to 37 per cent; the rest of the fish were salmon parr, three-spined stickleback, minnow and stone loach.[22]

The rates of growth of the fish were higher in the hard water streams than in the soft water stream. In all cases, growth was most rapid in spring and summer, when the water was warming up and plant life was developing most rapidly. There was also some growth in the winter, but it was much less than in the spring and summer.

In the hard water streams, the dominant sculpin showed a rate of production of 6.2 to 43.1 gm/m^2/fresh weight, and trout production 2.6 to 12.9 gm/m^2. Bullheads were not present in the soft water stream and trout production was at the rate of 12.1 gm/m^2. In these streams, the hard water ones were more productive, and of a greater variety of fish than the soft water streams.

NOTES

1. Odum, H.T. 1957. 'Primary production measurements in eleven Florida springs, and a marine turtle-grass community.' *Limn. Ocean. 2.* 55-112.
2. Depasse, P. 'Monographie Piscicole de la Province Orientale.' *Bull. Agric. Congo Belge*, XLVII (4) 1956.
3. Braun, R. 'Limnologische Untersuchungen an einigen Seen im Amazonasgebeit.' *Schweiz. Z. Hydrol.* XVI 1952.
4. Egglishaw, H.J. and Shackley, P.E. 'An Experiment on faster growth of Salmon, *Salmo salar* (L) in a Scottish stream.' *J. Fish. Biol.* (1973) 5: 197-204.
5. Chutter, F.M. 1970. 'Hydrological Studies in the Catchment of the Vaal Dam.' *Pt I. Int. Rev. ges. Hydrobiol.* 55 (3): 445-94.
6. Young, A. 1969. 'Present Rate of Land Erosion.' *Nature* 224: 851-2.
7. Egglishaw, H.J. 1964. 'The Distributional relationship between the bottom fauna and plant detritus in streams.' *J. Anim. Ecol.* 33: 463-76.
8. Egglishaw, H.J. 1969. 'The distribution of benthic invertebrates on substrates in fast-flowing streams.' *J. Anim. Ecol.* 38: 19-33.
9. Surber, E.W. 1933. 'Rainbow Trout and bottom fauna production in one mile of stream.' *Trans. Amer. Fish. Soc.* 66: 193-202.
10. Elliott, J.M. 1970. 'Diel changes in invertebrate drift and the food of Trout.' *J. Fish. Biol.* 2: 161-5.
11. Elliot, J.M. 1971. 'Life Histories and drifting of three species of Trichoptera.' *Oikos* 22: 56-61.
12. Elliott, J.M. 1971. 'Up stream movements of benthic invertebrates in a Lake District stream.' *J. Anim. Ecol.* 40: 235-52.
13. Hultin, L., Svensson, B., and Ulfstrand, S. 1969. 'Upstream movement of insects in a small Swedish stream.' *Oikos* 20: 553-7.
14. Hall, Charles. 1972. 'Migration and metabolism in a temperate stream ecosystem.' *Ecology* 53 (4): 585-604.
15. Stott, B. 1964. 'Movements of coarse fish in rivers.' *Nature* 190: 737-8.
16. Matheson, O.A. 1972. 'Enrichment of sockeye salmon lakes.' *Verh. Internat. Verein. Limnol.* 18: 1089-95.
17. Vaas, K.F. 1948. 'Notes on freshwater fish culture in domestic sewage in the tropics.' *Landbouw, Eatavia* 20. 1948.
18. Allen, R.H., Herbert, D.W.M., and Alabaster, J.S. 1958. 'A field and laboratory investigation of fish in a sewage effluent.' *Fish*, Invest. Ser. I. Vol. VI. No. 2. HMSO London 1958.
19. Ladle, M.W. 1971. 'The biology of oligochaetes from Dorset stream.' *Freshw. Biol.* 1: 83-97.
20. Kennedy, C.R. 1969. 'Tubificid oligochates as food of Dace, *Leuciscus leuciscis* (L).' *J. Fish. Biol.* 1: 11-5.
21. Egglishaw, H.J. 1970. 'Production of salmon and trout in a stream in Scotland.' *J. Fish. Biol.* 2: 117-36.
22. Mann, R.H.K. 1971. 'The populations, growth and production of fish in four small streams.' *J. Anim. Acol.* 40: 155-90.

Chapter 3 RIVERS

The course of a river can be arbitrarily divided into seven zones.[1] At their headwaters, as hill streams, they are in the 'torrent zone' with a typical slope of 1:4. In such fast-running water the aufwuchs flourishes, clinging, with its abundant infauna, to solid holds in the stream and in the splash zones of cataracts and waterfalls. Many insect larvae are able to flourish in such conditions, but not fish. But the next zone, the 'trout zone,' has an average gradient of about 1:50, still a fast flow but divided into riffles and pools. This zone has been dealt with in Chapter 2.

Lower down the valley comes the 'salmon zone,' with a gradient of 1:100. The stream will now include long quiet and deeper stretches where salmon can rest on their spawning migration. They are able to live for months on their stored food reserves and, of course, they cannot take food at this stage of their life history. Salmon, big heavy fish, are a fine sight lying quietly in some pool, keeping station in the swirling water and just rising from time to time to look at some bit of flotsam drifting downstream.

The next river zone is the 'grayling-dace zone,' where the fall is about 1:250, still steep enough to give a fair rate of flow; somewhere about this point the stream becomes in popular estimation a river. In the next two zones, the 'roach-rudd,' with a gradient of 1:600, and the 'tench-bream zone,' with a gradient of 1:2000, the current has become slow or hardly perceptible, and this is surely a typical river. The last zone, the 'flounder zone,' brings us to the estuary, which will be dealt with in Chapter 10.

Primary Production in Rivers

Rivers are productive bodies of water, and their basic production can spring from plankton, from the bottom vegetation, from detritus produced in the river itself, from detritus carried or blown into the river as leaves, organic pollutants, or from terrestrial insects (a more important source of food material than might be suspected).

A study of primary production has been made in the River Thames at Reading, in the lower part of its course in the roach-rudd and tench-bream zones, though the river is canalised and flows over a series of weirs, with almost still water between.[2] Where the study was made, the Thames is joined by a tributary, the Kennett, which usefully contrasts with the main river. The flow of the river, between the cascades over the weirs, is slow and the resident time is about 6 days; it was therefore surprising to find a rich phytoplankton of truly planktonic forms, and not benthic forms carried in suspension. In the Thames above the Kennett confluence, phytoplankton accounted for 75 per cent of the total organic matter in suspension in the month of June, decreasing to 2 per cent on November. The mean annual dry weight of organic matter in suspension was 7.05 gm/m^3, of which 2.03 gm/m^3 (28.8 per cent) was phytoplankton. By contrast, in the Kennett, the sewage outfall of the city of Reading, 4 km. upstream, resulted in an annual mean dry weight of 7.84 gm/m^3, but only 5.1 per cent of this was phytoplankton.

Above the Kennett confluence, plankton community photosynthesis greatly exceeded respiration, leaving a net positive balance of 4,388 $kcal/m^2/year$. In the Kennett, there was an excess of respiration over photosynthesis, giving a net community deficit of -275 $kcal/m^2/year$. But as the volume of the Thames water is normally much greater than that of the Kennett, below the confluence the river has a positive net community production of 1,907 $kcal/m^2/annum$. The liberation of abundant nutrients by the mineralised sewage solids leads to the high planktonic production in the main river.

This planktonic production was supplemented by minor contributions from macrophytes (willow trees, water lilies, reeds) and by epibenthic algae; the latter were not estimated, but were believed to be much less than that of the phytoplankton.

Truly planktonic animals are not common, and cladocera and rotifers belong to forms which live in dense clumps of vegetation or in coarse gravel. They feed on the phytoplankton, and are fed upon by young fish. But the most important producers among invertebrates are the filter feeders, bivalves, sponges, and a Bryozoan. This benthos is under very heavy grazing pressure from the fish; indeed, the sponges are always in tatters, and though the biomass is at least 100 $kcal/m^2$ it is insufficient to support the fish population. The larger fish have to rely on terrestrial insects taken at the water surface and on organic detritus, even though this material is not very digestible.

The growth rates and fecundity of the principal fish, roach (*Rutilus rutilus*) and bleak (*Alburnus alburnus*) are among the lowest in any river, and are attributable to a dense population (100 seinable fish per 100 m^2, or about one million per 2 km. of the river) in relation to the available food supply. Production is very impressive, at about 200 $kcal/m^2$, or 2000 kg/ha. About 70 per cent of these are attributable to young fish in their first year of life and are not realisable as yield. It is suggested that, by management, a realisable harvest of about 500 kg/ha a year could be obtained, if an economic use could be found for it. There is no market for such fish in England. But fishing competitions are frequently held and the fish are returned to the river alive. So there is little fishing mortality, and angling clubs seem to prefer a policy of a high population even if the fish are small, so that anglers at least have a good prospect of catching something for their day's sport.

Studies have also been made on the River Stour, in Suffolk, a river familiar through the paintings of Constable and his school. The Stour flows through flat country, and is almost completely canalised. For most of the year its flow is scarcely perceptible. By contrast, the River Severn, one of the largest English rivers, has a fast flow in its upper reaches, but a slower flow in its middle and lower reaches.[3] In both these rivers phytoplankton was plentiful, especially the centric diatoms known as *Cyclotella*; but was most abundant in the slow flowing Stour and lower reaches of the Severn. In the Stour, the summer populations of the diatoms may be as high as 48 to 87 thousands of cells/100 ml.; and even at times of irregular flow there could be 37,000 cells/100 ml., of which 30,000 would be centric diatoms.

In the upper reaches of the Severn at Preston Mountford, where the river has a fast flow, diatom totals seldom exceed 10,000 cells/100 ml., even in summer. But at Upton, in the middle reaches, it is usual to find a count of 20,000 to 30,000 cells/100 ml.

Nanoplankton and Fixed Plants

But in addition to this production by the so-called net plankton, that is, phytoplanktonts large enough to be caught in finemeshed nets, there is an equally important primary production by the nanoplankton, a group of organisms measured in thousandths of a millimetre, which can pass through the meshes of the finest plankton nets and in many cases live either as plants, fixing dissolved carbon dioxide, or as animals, engulfing other minute prey such as bacterial cells. It was estimated that, in the River Stour, this nanoplankton contributed the substantial production of 0.57 mg./litre, dry weight.

Besides the phytoplankton and nanoplankton, fixed plants such as reeds and grasses, rushes and water lilies, and true water plants such as the pondweeds contribute to the primary production of the river. In some South American rivers, where primary production by the phytoplankton may be as low as 100mg.C/m^2/day, there may be a high rate of production due to floating meadows of the grass *Paspalum repens* and species of *Eichhornia*, the water hyacinth. Besides their own production, such water meadows harbour a moderate aufwuchs, which can start a food chain.[4]

Allochthonous Sources of Material

Production by phytoplankton and by fixed vegetation is autochthonous production, produced in and by the river. But rivers also receive great amounts of organic matter produced from the land, which can be called allochthonous. In temperate climates, the most important is the leaf fall. In the Thames at Reading, this annual input of leaf litter has been estimated at 1.6 kg./metre of shoreline, or 23.2 gm/m.^{2}/year.[5] Observations on the natural breakdown of this material showed that the most important agents concerned were bacteria and fungi. Larger invertebrates, though they colonised the litter in large numbers, did not break it down faster than the micro-organisms alone. As with organic detritus generally, the particles of detritus, as they are broken down into progressively smaller fragments, become enriched in nitrogen, and thus become a more valuable foodstuff.

In spite of this, it was estimated that the energy input from leaf litter was only a small proportion of the metabolism of the fish population of the river, though its nitrogen content could be an important factor in the economy of the river.

Downstream Drift of Benthic Organisms

Estimates of the bottom fauna of rivers can be inaccurate if they are based solely on sampling by grab or dredge. In the Amur River of eastern Siberia, the benthic fauna, as judged by grab and dredge samples, was believed to be sparse, until the seston, or drifting organisms of the river, was sampled by netting. Then it was found that a large part of the bottom fauna takes part in this drift, especially at the time of summer floods due to snow meltwater.[6] The drift was then found to include not only the larvae of stoneflies and mayflies, but also oligochaete worms, leeches, molluscs and freshwater prawns. There were also ants, cicadas and springtails obviously washed into the river from the banks or from the air. This drift, as was found also in the streams (Chapter 2), is not wholly mechanical and due to the violence of the flood water, but purposeful and intentional, since the drift is far more active at night, when also there is a well-marked vertical migration from the bottom into midwater.

Primary Production in Tropical Rivers

In tropical rivers, the phytoplankton is the most important source of primary production. In the Blue Nile,[7] as also in the Severn and Thames, the concentration of phytoplankton is low during the flood season, when there is rapid flow and great turbidity. But with the subsidence of the floods, there is a great eflorescence of the phytoplankton, and especially of diatoms such as *Melosira*. The still waters of lagoons and backwaters, and of reservoirs, favour the rapid growth of diatoms even at the time of flood, and these are the sourcc of the seeding of the main river. They undergo great development when the silt settles enough to allow a deep penetration of the sunlight. In the Nile, this outgrowth of the phytoplankton is checked by the exhaustion of the available supplies of nitrate-nitrogen. Then there can follow a high density of the blue-green algae such as *Anabaena flos-aquae*, which can cause a high supersaturation of oxygen and a depletion of phosphate.

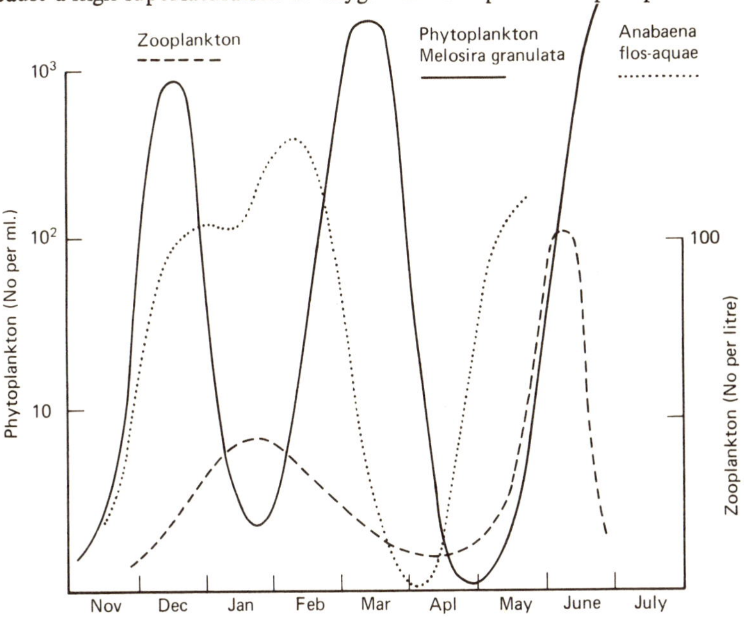

Figure 4 Succession of a diatom, a blue-green alga, and zooplankton in the Blue Nile. Modified from Talling and Rzoska.[7]

A rich zooplankton develops to feed on this bountiful phytoplankton, consisting chiefly of crustacea such as ostracods, daphnids and cyclopoid copepods. The massive grazing of these filter feeders can bring about a sudden fall in the numbers of the phytoplankton, as is also the case in the English Stour. Many fish and insects feed on the zooplankton, and form a food-chain which distributes the primary production throughout the river.

The seasonal regime of the River Sokoto in northern Nigeria is also dominated by the seasonal flooding.[8]

Stratification

Rivers and streams are usually well mixed and well oxygenated by natural turbulence, or by falling over manmade weirs. This is so usual that limnologists are content to take a single observation for temperature, chemical composition and oxygen content, assuming that this one sample will represent the whole water body. But when rivers become deep and slow, this may no longer be the case. For example, some of the tributaries of the River Amazon, big rivers in their own right, such as the Rio Tapajos, may be 30 metres deep and flow so slowly, at about 200 metres per hour, that they are almost stagnant. In such a river, mixing may no longer occur, and the surface layer, heated by the sun, may come to lie over heavier cooler water, with a sharp discontinuity layer at a depth of two to four metres. This discontinuity layer effectively isolates the deeper water from the atmosphere; and in such conditions the deeper water, in contact with the rich organic deposits on the bottom mud, can become deoxygenated and unable to support life. This phenomenon of stratification becomes very important in large bodies of water such as lakes, and of course in the great ocean itself (Fig. 6).

River Fish and Fisheries

Though rivers are productive bodies of water, the greatest river fisheries are for migrant fish such as salmon, hilsa and eels.

A salmon weir is a barrier built across a river, with a number of gaps left in it through which the fish must ascend on their struggle upstream. Traps can be placed in these gaps into which the salmon collect, to be gaffed or netted out. Salmon weirs in rivers are now uncommon, but they were the principal means of catching salmon in those far off days when salmon was so plentiful that it was a poor man's food. Dutch paintings frequently show salmon on the tables of humble people, but few modern Dutch folk have seen a salmon outside a can or deep freeze. The gross pollution of the Rhine has seen to that.

Salmon fisheries in rivers now tend to be sport fisheries, and are very valuable properties in Europe, eastern Canada and in western North America, especially for king and coho salmon. The various kinds of trout are in the same class.

Shads

The shads are also fish which ascend rivers to breed; one of the most important is the hilsa (*Hilsa ilisha*) of the main rivers of India, and of the Indus. Medium-sized fish, they weigh 3 to 5 lbs. (1.5 to 2.3 kg.). They ascend for hundreds of miles, crossing obstructions such as irrigation weirs. They are in prime condition, fat and nutritious and full of roe, itself a delicacy. No wonder the annual migration is eagerly awaited. Fishing camps are set up, where the fish are caught and cured for the market, and thousands of fishermen set gillnets in the rivers and also use many types of hand and scoop nets.

The bonga of west Africa (*Ethmalosa fimbriata*) is another of the shads, which ascends the estuaries and lower rivers in the autumn, at the end of the rains, and provides the most important of the estuarine and river fisheries there. Canoes loaded with fishing gear, hut making materials and domestic supplies may be seen, at the beginning of the season, bound for the fishing areas, where camps will be established and a busy trade carried on.

Eel Fisheries

There is a year round fishery for the yellow eels, the growing immature fish, which move about considerably in the summer months and are generally migrating upstream. When the elvers enter freshwater after their long drift across the ocean, those which remain in the estuaries and the lower reaches tend to be, or to become, males, which are smaller than the females. Those which migrate further upstream grow larger and are, or become, females. All through their long lives in fresh water, they tend to move further upstream, though recent work in Ireland has shown that this movement may be very slow, so slow that the upper reaches of the river could probably carry a much larger stock of eels than they do. Probably it was this consideration which prompted the German, and now the Polish, fisheries departments to import live elvers from the elver fishery in the estuary of the English River Severn, to release in their plentiful freshwater areas.

But the biggest eel fisheries are those for the adult and maturing eels, which take on a silvery hue and have enlarged eyes, both modifications which anticipate the change of habitat from life in a tranquil river far inland to a short but intense life in the deep ocean. Naturally, these fish are in prime condition, and, when well-smoked, make one of the world's delicacies. As they migrate downstream, they are caught in barriers similar to the salmon weirs mentioned above. A famous one is the eel weir at Toome, on the River Bann in Ireland, which drains Lough Neagh, the largest body of inland water in the British Isles. It is a permanent structure of concrete, spanning the river but with many gaps which in the season are closed by nets, and catch very large numbers of silver eels on migration. Much bigger eel fisheries occur in the Isselmeer in Holland, where the mature eels seeking a way to the sea are headed off by long barriers of netting and steered into non-return cages of netting. Greatest of all is the eel fishery in the western

Baltic, where fish fences similar to those in the Isselmeer occur for scores of miles along the coasts, and especially along the coasts of Denmark. These traps are in a position to take the eels which have grown in the watershed of the Baltic; no wonder that Denmark supplies most of Europe with eels, and especially smoked eels.

There are minor eel fisheries associated with water mills. A barrier prevents them from passing over the water wheel and guides them to a 'fish house' with a slatted floor, through which the water falls, leaving the eels stranded on the slats. In old deeds, a part of the rent of the mill would be in the form of a stipulated number of eels.

Eel fisheries were also carried on by the Maoris in New Zealand, and for the marbled eel, *Anguilla nebulosa*, in India and Kenya. High up on Mount Kenya, migrant Kikuyu fisherman would come in the season to trap the big full grown eels as they came down the mountain streams. They are also found in the new Lake Kariba (Chapter 6).

There are many locally important fisheries for fish ascending rivers from lakes. Some of those in Africa have been described.[9]

Predatory fish seem to predominate in the river fisheries of Europe and North America, but in tropical river fisheries a high proportion of the fish caught are herbivores, such as the *Labeos* and *Tilapias*, and insectivores, such as Barbus species. Important detritus feeders, such as *Citharinus* and *Distichodus*, are a valuable part of tropical river fisheries. In large rivers, at the time of low water, as in the Nile and Volta, for instance, teams of migrant fishermen work gillnets and seines. These are not great fisheries, as are the fisheries for migrating eels, but they are locally valuable. Large exports of dried fish result from the seine fishery in the Sudanese Nile.

Floods and Flood Fisheries

Rivers change their courses over the centuries, and abandoned reaches of the river are called 'ox-bows' which become seasonal or perennial lakes. Rivers may also carry a large load of suspended silt, for the water is always wearing the land away. Some of this will settle where the force of the stream is slackened, as along the banks, where friction slows down the current. Tropical rivers, unlike those of temperate climates, do not usually flow in a definite 'thalweg,' and it is common to find that the banks of such rivers are higher than the surrounding valley, and that when the river overflows at the time of annual or occasional floods, there is widespread flooding. The silt may be deposited in the valley, and fertilise the crops which will be taken as the floods recede. Thus, for example, the inhabitants of the Zambesi valley took their annual crops of maize and millet.

When the rivers overflow their banks, and create large areas of shallow water, many or most of the river fish follow, and find in the flooded land very rich feeding. In place of the limited resources of the river bed, the fish now have access to large areas abounding in insect and invertebrate life of all kinds, and to living and rotting vegetation, which becomes covered with aufwuchs. The flooding is also the trigger for spawning, so that, during the flood, which may last for four to six months, not only do the adult fish feed and grow well, but their off-spring get a head start in life and may be close to commercial size themselves by the time they regain the river. Many traps are set to catch these fish on their way back to the river. In the Zambesi flood plain in Barotseland, a survey showed that there were an average of five active fish weirs per square mile or a total of 10,000 fish weirs in the flood plain.[10] Yet the fishery still gives its annual harvest of first class protein decade after decade; so clearly enough fish survive to keep up the stock of fish, and the survey wisely decided not to meddle. Many fish are trapped in drying lagoons as the flood water evaporates, and it is clearly an economy to take all such fish if possible, as they will die in any event. But in southern Russia there is an annual salvage of trapped fish of economic importance, for restocking in rivers and ponds.

Some very large fish crops are got from flood fisheries. In the Kapua River in Indonesian Borneo the shaols of *Thynnichthys* which have fed in the flood forests give an annual crop of 17,500 tons.[11] Though caught in the rivers they are the product of the flood forests. In the far less important flood fisheries of the upper Baram, in Borneo (Sabah), large catches are made in liftnets. The greatest of all the flood fisheries is that in the flood forests of the Mekong River in
in the water, and plant plankton production is very low, at an estimated 100 mg.C/m^2/day. Yet these rivers support a fair sized fish population, as the fish have access to the land vegetation in the flood season. These floods can be very high and the floods are correspondingly wide. The Orinoco rises by fifteen metres, the Rio Negro by ten metres, and even the Amazon itself rises by eight metres at Santarem.

Some very large fish crops are got from flood fisheries. In the Kapua River in Indonesian Borneo the shoals of *Thynnichthys* which have fed in the flood forests give an annual crop of 17,500 tons.[11] Though caught in the rivers they are the product of the flood forests. In the far less important flood fisheries of the upper Baram, in Borneo (Sabah), large catches are made in liftnets. The greatest of all the flood fisheries is that in the flood forests of the Mekong River in Cambodia, which gives, or rather gave, an annual fish crop of 100,000 tons. In the middle Niger, a fishery for *Alestes* gives an annual crop of 45,000 tons, greater than that of the sea fisheries.[12]

River Barrages and Dams

The natural flood regime of the rivers is seriously upset by the building of barriers and dams. The barrage now being built across the Mekong may ruin the flood fishery there, a high price to pay for an irrigation scheme. Such dams are designed to even out variations in the river flow, and thus abolish the annual floods, with their natural irrigation and large fish crops. Below the new Kainji Dam on the River Niger, the annual floods, which were in two periods, the 'black flood' with its peak about February-March, and the 'White Flood' with a well-marked peak in October, changed into irregular fluctuations of level of several feet – for instance, changes of three feet in two or three days.[13] There was a serious decline in the catches of fish, and a fear that, now the spawning trigger of the regular floods has gone, the breeding of the river fish may be affected. Before the closing of the Kariba Dam in Zambia the annual floods watered a rich pasture of grasses and shrubs which attracted and nourished large herds of game.[14] Now there is no annual flood, and the valley is taking on a semi-desert character. The game herds must be reduced to cope with the less productive conditions and drought-tolerant species will be favoured. The new dam across the Senegal River in West Africa has had a profound effect on both the agriculture and the fisheries in the river basin below.

Further, regular floods are, for many fish, the trigger for spawning. If these floods are controlled by a barrage so that, instead of occurring regularly, they can be made to occur at any time and in any volume, they can have an important effect on the breeding of river fish. It has been shown that the salmon will not ascend the river to spawn except in response to a flow of new water. Irrespective of the actual volume of the water, whether low flows in summer, or high flows in spring and autumn, the salmon respond to sudden freshets.[15] This can indeed be used to manipulate the upstream runs of salmon. In India, the discharge of water from the Hirakud Dam on the Mahanadi River in Orissa State could cause a temporary rise in the river level and an increase in the speed of flow. The river fish would respond by spawning in the rising and fast-flowing shallows. The fry fishermen come by scores and set their small-meshed filter nets on stakes driven into the river bed, and take hundreds of thousands of fry per net in an hour or two. There is a good market for the living fry of the major Indian carps in Calcutta.

The suggestion that the White Nile might be carried by a canal across the swamps of the southern Sudan, thus abolishing the annual flooding of the 'sudd,' had to take into account not only the loss of pasture for the Nilotic tribes, but the effects on the changed breeding regimes of the fishes of the area.

NOTES

1. Fort, R.S., and Brayshaw, J.D. 1961. *Fishery Management*. London, Faber & Faber.
2. Mann, K.H. 1972. 'Case-History: The River Thames,' in *River Ecology and Man*. New York and London. Academic Press. pp. 215-32.
3. Swale, E.M.F. 1969. 'Phytoplankton in two English Rivers.' *J. Ecol.* 57: 1-23.
4. Gessner, F. 1964. 'The Limnology of tropical rivers.' *Verh. int. Verein. theor. angew. Limnol.* 15: 1090-1.
5. Matthews, C.P., and Kowalczewski, A. 1969. 'The disappearance of leaf-litter and its contribution to production in the Thames.' *J. Ecol.* 57: 543-52.
6. Klyuchareva, O.A. 1963. 'On drift and diel vertical migrations of bottom invertebrates of the Amur.' *Zool. Zh.* 42: 1601-12. (In Russian.)
7. Talling, J.F., and Rzoska, J. 1967. 'The development of plankton in relation to hydrological regimes in the Blue Nile.' *J. Ecol.* 55: 37-662.
8. Holden, M.J., and Green, J. 1960. 'The hydrology and Plankton of the River Sokoto.' *J. Anim. Ecol.* 29: 65-84.
9. Hickling, C.F. 1971. *Fish Culture*. London. Faber & Faber.
10. Bell-Cross, G. 1971. 'Weir fishing on the Central Barotse Flood Plain in Zambia.' *Fish. Res. Bull. Zambia.* 5: 331-40.
11. Saanin, J. 1952. 'Notes on the fishery of the Cyprinid fish *Thynnichthys vaillenti* in Indonesia.' 4th Meeting Indo-Pacif. Fish. Coun. Prrc. Quezon City. F.A.O. Bangkok.
12. Blanc, M., Daget, J., and d'Aubenton, F. 1955. 'L'Exploitation des eaux douces dans le bassin du moyen Niger.' *Bull. I.F.A.N.* XVII Ser A. (4) 1955.
13. Lelek, A. 1972. *Kainji Lake Research Project*. F.A.O. 1972.
14. Attewell, R.I.G. 1970. 'Some effects of Lake Kariba on the ecology of a flood plain of Rhodesia.' *Biol. Conserv.* 2: 189-96.
15. Alabaster, J.S. 1970. 'River flow and upstream movement and catch of migrating salmon.' *J. Fish. Biol.* 2: 1-13.

Chapter 4 LAKES (I)

A lake could be defined as a body of water in which the resident time is long. Even in the very slow-moving Amazon tributaries mentioned in the last Chapter, or the canalised rivers, the resident period of the water is short. But in Lake Victoria, for example, the resident period of the water is over a century, in Lake Baikal several centuries, and in Lake Tanganyika thousands of years.[1] This has biological significance, for when water remains in one place for a long time, it takes on properties different from those of flowing waters; we have a lotic, as distinct from a lentic, environment. Because of this, and their large area, lakes are important sources of fish.

Lakes may be formed in several ways. Some are water-filled depressions caused by earth movements, by volcanic activity; some are formed by landslides and flows of lava, which block the exits of valleys, which then fill with water. Some are formed by the gouging action of millions of tons of ice, slowly crushing their way to the sea in glaciers, and depositing dams in the form of moraines of rock and soil.

Lakes may differ from each other hardly less than other areas of land. The geological history and the nature of its watersheds affect the character of a lake, and of the flora and fauna which it will support. Lakes whose basins bring in plant nutrients in quantities large enough to support a rich plant, and thence animal, life are called eutrophic; while lakes in basins of granite or other insoluble rocks are poor in plant and animal life and are called oligotrophic. Lakes due to the action of glaciers in primitive rocks, such as all the thousands of lakes in the Canadian shield region, and the Great Lakes of North America, are, or originally were, oligotrophic. The prevalence of winds which blow over the lake surface, the elevation above sea level, the latitude and consequently the amount and seasonal nature of heat received from the sun and the depth of the lake all affect its character.

The most celebrated of the lakes formed by massive earth movements are Lakes Baikal and Tanganyika, which offer a complete contrast to each other. Both are the deepest, the most ancient, and among the largest lakes of the world. Both are ancient enough to have evolved specialised faunas of animals endemic to them alone; Baikal is subarctic, oligotrophic; Tanganyika eutrophic and tropical. Baikal is well oxygenated to the bottom at over 1700 metres; Tanganyika has no oxygen at depths greater than about 200 metres, so that most of the great volume of the lake is lifeless.

The Life History and Ageing of Lakes

Because there are long-term fluctuations of rainfall and climate, lakes may change their levels and volumes over the millennia, and most have done so, as revealed by raised (and submerged) terraces which mark former lake shores. Some lakes now separate were formerly joined. But generally speaking lakes are ephemeral bodies in the geological sense. Material gradually accumulates in the lake, making it shallower; windblown soil, dead plankton, detritus derived from plant beds, soils and detritus carried in by affluent streams and rivers; these are among the causes of the progressive infilling of a lake. As the lake becomes shallower, fringing aquatic vegetation (reed beds, papyrus and the like) encroaches into the open water and in turn deposits vegetable matter.[2] The lake gradually turns into a swamp, in which open water occurs as patches between fields of fixed and floating vegetation. Lake Chad, in north eastern Nigeria, still has an area of 7,000 square miles, but the eastern half is full of swampy low islands, and the lake has lost one third of its area since the mid-nineteenth century. The great accumulation of organic matter uses up the dissolved oxygen; the wind has no chance to stir in fresh oxygen; an acid environment is created; and there may be an accumulated 'oxygen debt' in the form of fermentation products, and the water will not support life until the debt is paid off. Most of the inhabitants of such swamps have devices enabling them to be independent of dissolved oxygen; the fish, for instance, have many and various air breathing organs supplementary to their gills. In such swamps in Africa, the herds of hippos play a valuable part in consuming great quantities of vegetation and helping to keep some open water. Nevertheless, some swamps are important producers of fish, especially in times of flood, when, as with rivers, the fish may spread out over the newly flooded land, finding good feeding conditions, and being caught in traps and barriers as they regain the permanent swamp at the end of the floods.[3]

This infilling of a lake can continue until the lake becomes a bog, and from thence dry land. In small lakes this infilling can take place in one man's lifetime; but it takes thousands of years with larger lakes. Where a lake has a river discharge, erosion of the sill during millennia will ultimately drain the lake dry. At the same time, new lakes are being formed. Because of the ephemeral nature of most lakes, their flora and fauna is limited, and tends to be the same over most of the earth.

There are also many temporary lakes in desert lands which fill when the rare rainfall comes. Such are many Australian lakes, like Lake Eyre which is a large body of water at intervals of several years;

these evaporate into salt pans, plain to see in flights over the 'outback.'

Very large lakes are able to create their own climate, and this includes very heavy rainfall directly on to the lake surface; such lakes have been well described as enormous dewponds. This direct rainfall may be the most important source of the water in the lake; it brings in little material aside from some nitrogen compounds derived from lightning discharges in the atmosphere and from the smoke of forest fires. The rainfall in the catchment may contribute less water, but contributes dissolved salts which accumulate in the water as the result of evaporation.

Bahi Swamp in Tanzania fills in most wet seasons, and a brief and curious fishery for catfish occupies hundreds of migrant fishermen as the waters evaporate to hardbaked mud.[3]

Physical Conditions in Lakes

The latitude and elevation of a lake affect the temperature of the water, and hence the speed and range of the natural processes which take place there. It is found that the water at the bottom of lakes has about the temperature of the lowest ambient air temperature of the year. This is because cold water is heavier than warm water, and the heaviest water will be that cooled at the coldest time of year locally. Thus, in temperate climates the bottom temperature of lakes is about 4°C, as water at this temperature has a maximum specific gravity; while in the tropics it will depend on altitude. In a series of Indonesian lakes at near sea level, the bottom temperature was about 26°C, but in Lake Toba in the same area, which lies at 730 metres, it was 24.6°C, and was 20.2 in a lake at 1,531 metres. In tropical Lake Titicaca, at 3,800 metres, the bottom temperature was 11°C.[4]

Stratification of Lakes

The sun warms the surface water of lakes; but water is a poor conductor of radiation, so that the direct effect of the sun's heat is confined to the upper few metres of the surface. Warm water is lighter than cool water; so a condition of stability can be set up, in which the lighter warm water gives the water column a high centre of gravity. But where wind is free to blow over a lake, it drives the surface water before it, and by setting up turbulence, currents and countercurrents, it causes the warm surface water to penetrate deeper, the energy of the wind providing the work needed to overturn the stable water column. The transition between the warmer surface layer and the deeper cold water may become abrupt, a considerable fall in temperature taking place in a very few metres; this discontinuity is called the thermocline. The surface waters above it are called the epilimnion, the deeper waters below it, the hypolimnion. Generally, the thermocline occurs at a greater depth in larger lakes than in smaller ones. In small Indonesian lakes, the thermocline occurs at four to eight metres; in larger lakes of about 100 $km.^2$ area, at about twelve to fifteen metres; in Lake Toba, of 1,000 $km.^2$, at 30 metres; in Lake Titicaca, of 7,700 $km.^2$, at 50 to 100 metres according to season, and in lakes Tanganyika and Malawi, of over 15,000 $km.^2$, at 150 to 200 metres. The reason for this relation between the surface area of a lake and the depth of the thermocline is that the larger the area over which the winds can blow freely, the greater their power to mix, and drive deep, the warmed surface waters of the lake.

The thermocline is an effective barrier against the mixing of the waters of the epilimnion and the hypolimnion, the waters above and below the thermocline respectively. The epilimnion is also the layer which receives the incident sunlight and in this layer the phytoplankton develops until it may use up all the nutrient salts there; then production must stop, even though, in the hypolimnion, there may be an abundance of unused nutrient salts which, however, cannot diffuse up through the thermocline. The epilimnion may become saturated with dissolved oxygen, while, below the thermocline, the lack of photosynthesis, together with the fall of organic matter, may cause an anaerobic hypolimnion in which nothing can live. An outstanding case is the eutrophic Lake Tanganyika; whereas in the oligotrophic Lake Baikal photosynthesis proceeds in the very clear water to a depth of sixty to seventy metres, and the whole water column is fully oxygenated.[5]

Chemical Stratification

There is another kind of stratification in lakes, namely, that due to their chemical composition. Where water contains dissolved salts, it will be denser than pure water, and may remain denser even when warmer. There can then be an inverse stratification. Lake Kivu, in the Rift Valley of East Africa, was formed a mere 15,000 years ago by the eruption of one of the volcanoes which are still active today; a lava flow dammed the valley of a river which flowed north into the Nile watershed, and, when it filled with water, it now overflowed south to the watershed of the Congo. Many hot and highly mineralised springs discharge into its deeper waters. So in this lake, cool purer water overlies warm salt-bearing water. A thermocline exists, but 99 per cent of the volume of the lake contains no dissolved oxygen and can support no aerobic life.[6] This deep water of Lake Kivu is saturated with methane at an estimated total volume of 50 milliards of cubic metres, which bubbles out of solution when water samples are brought to the surface. A decade ago, a bold scheme was put forward in which this deep water, which is also rich in nitrogen and phosphorus plant nutrients, should be pumped to the surface. The methane could be collected and used and the deep water, now allowed to warm in the atmosphere, would be spread over the lake surface, where its rich load of nutrient salts would nourish such a bountiful harvest of phytoplankton that the annual fish production of the lake could be raised from the existing 1,000 tons to a possible 35,000 tons;[7] the more easily if the attempt to transplant freshwater and planton-eating sprats from Lake Tanganyika were successful, as it has been in Lake Kariba (Chapter 6).

Inverse Stratification

Inverse stratification is also shown in some notable freshwater lakes in Antarctica.[8] Though perpetually covered with twelve feet of ice, enough sunlight penetrates into these very clear waters in the southern summer to warm the under-ice water layer to 5°C. With increasing depth, the water gradually grows warmer until the temperature at twenty metres is 9°C. Then there is a sudden increase in temperature to 24°C at fifty metres and thence to the lake bottom; this must surely be the warmest place in the whole of Antarctica. The water of these lakes is so clear that light penetrates to the bottom at seventy metres. This heating of the deep water is due to its higher salt content, which prevents water circulation; a 'greenhouse effect.' Life in these lakes is affected by these unusual conditions. In the deep warm water, plants, chiefly flagellates, develop plentifully near the bottom, and bacteria are most abundant at the inverse thermocline at twenty metres. Flagellates predominate because, in the absence of turbulence, organisms must be able to counteract the tendency to sink. These flagellates can produce new organic material during the continuous daylight of the southern summer, by photosynthesis; while in the perpetual darkness of the southern winter they can live on dissolved and particulate organic matter in the water. On a summer day, production was measured at 29 $mg.C\ m^{-2}\ day^{-1}$. The availability of nutrient salts could be a limiting factor to the production of these plants, as well as daylight and temperature; for the addition of 1 p.p.m. of nitrate nitrogen to a sample of lake water increased carbon assimilation by 41 per cent over controls without added nitrogen.

Shallow Lakes

Stratification of the water is not likely to occur in wide and shallow lakes, where the water will be completely mixed from top to bottom by the wind. Such are Lakes Chilwa, in Malawi, and Rukwa, in Tanzania. In Chilwa there is a high rate of fish production and a prosperous fishery; but from time to time exceptionally strong winds can so stir up the lake that the bottom silt, very rich in organic matter, comes into suspension and uses up all the oxygen in the water resulting in a massive fish-kill.[9] This shallow lake also dried out recently and the fishery ceased. Plans were made to breed artificially fish with which to restock the lake when it refilled. Lake Rukwa also dried out in 1949, and the world's press carried fine photos of herds of hippos dying of thirst, of fish rotting in heaps, and of the lake bed, now a steaming swamp, grown into reed beds and ambatch shrubs with their bright yellow flowers.

But, in both of these shallow lakes, it took only two years for the fish to make a complete recovery after the refilling, and that without the need for any restocking. In each case, a few fish must have survived in some pools or river estuaries, and their rate of reproduction in such favourable conditions of temperature and fertility was high enough to restore the high level of the fish biomass. I visited Rukwa in 1951, two years after the lake refilled, and measured many of the fish caught by gillnet. They were only slightly smaller than the fish in the lake before the drying out, but just as plentiful.

Overturns

Earlier in this Chapter it was shown that, in most lakes, a stratification sets in, such that a warmer water layer, the epilimnion, overlies a cooler hypolimnion, separated from it by a thermocline in which there is a rapid transition in temperature. But this stratification can be upset, either seasonally, or catastrophically. There is a regular overturn of the waters of lakes in temperate climates. In high summer, the surface temperature can be 15° to 20°C, while the bottom water may be about 4° to 5°C, with a marked thermocline. The nutrients in the epilimnion are soon used up by the growth of the phytoplankton and submerged marginal higher vegetation; while in the hypolimnion there may be unexhausted nutrients. In the autumn the surface layers cool down until the whole lake has a uniform temperature, and there is nothing to prevent a complete mixing of the nutrient-rich bottom waters with the exhausted surface waters. If at this time there is still enough sunlight, there may be a second efflorescence of plant life. But then winter comes, sometimes with ice cover, leaving the lake unstratified and homogeneous, as to temperature and chemical content, until the surface waters of the lake are warmed up again in the spring.

But in shallow tropical lakes, the stratification can be upset both seasonally and catastrophically. This is because, at temperatures of 28° to 32°C, the specific gravity of water changes rapidly and there can be a stable thermocline even though the difference between epilimnion and hypolimnion may be only a couple of degrees or less. This small temperature difference can be upset by only a few cold nights, and an overturn can take place. Where there had been a large anaerobic hypolimnion, such an overturn can result in the mixed water of the lake coming to have too low an oxygen content to support fish life, and a fish-kill results. There may also be a seasonal overturn in monsoonal climates, where clear nights in the dry season allow a rapid radiation of heat from the lake surface.

The deep Lake Malawi may have a partial overturn every couple of decades, though there are no accounts of massive fish-kills. The shallower southern part of the lake does have an annual overturn; and this brings enough of the nutrient-rich deeper water to the surface to nourish a productive plant life, which of course is the basis of all the fish life in the lake. Lake Tanganyika has only a partial overturn of the upper 150 metres, but it gives a fish harvest of about 80 kg./ha./year. Much higher fish crops are given by shallow tropical lakes which have no overturn. Besides Lakes Rukwa and Chilwa already mentioned, the small and shallow Lake Kitangiri in Tanzania produces at a rate of 165 kg./ha./year, and Nakivali, in Uganda, at 149 kg./ha./year. Lake George, also in Uganda, is productive, at about 93 kg./ha./year; but in this very shallow lake there are occasional fish-kills due to the stirring up of the bottom mud, as in Lake Chilwa. Incidentally, African and other fishermen imitate these natural fish-kills, when fishing small lakes and ponds; they trample the water until the fish come to the surface suffocated, when they can be caught by hand or basket.

Natural Reserves of Plant Nutrients in Lakes

In some permanently stratified lakes, large amounts of nutrient salts have accumulated in the hypolimnion. In one small Indonesian lake, only 750 metres in diameter, and 28 metres deep, the hypolimnion contained in solution 1,430 kg. of phosphorus, and 12,171 kg. of ammonium nitrogen. In a larger lake, also in Indonesia, with an area of 98 $km.^2$ and a depth of 170 metres, the hypolimnion contained 1,500 metric tones of phosphorus and 7,000 tons of ammonium nitrogen.[10] If irrigation water were to be drawn from the hypolimnion of lakes such as these, the irrigated crops would need no fertiliser.

NOTES

1. Beauchamp, R.S.A. 1964. 'The Rift Valley Lakes of Africa.' *Verh. Internat. Verein. Limnol.* XV: 91-9.
2. Ruttner, F. 1940. *Fundamentals of Limnology.* English translation. University of Toronto Press, 1953.
3. Hickling, C.F. 1961. *Tropical Inland Fisheries.* London. Longmans Green & Co.
4. Gilson, H.C. 1964. 'Lake Titicaca.' *Verh. Internat. Verein. Limnol.* XV: 112-27.
5. Moskalenko, B.K. and Votinsev, K.K. 1972. 'Biological productivity and balance of organic substance and energy in Lake Baikal.' *Proc. IBP-UNESCO.* Symposium on the productivity problems of freshwaters, Poland, May 1970: 207-25.
6. Damas, D. 'Recherches hydrologiques dans les lacs Kivu, Eduard, et Ndagala.' *Inst des Parcs Nat Congo Belge.* Mission H. Damas. Fasc 1 Bruxelles. 1935-6.
7. Capart, A. and Kufferath, J. 1956. 'Recherches hydrologiques au Congo Belge et leurs resultats pratiques.' *Bull. Agric. Congo Belge.* XLVII (4) 1956.
8. Goldman, C.R. and Hobbie, J.E. 1967. 'Two Antarctic Desert Lakes.' *Limn. Ocean.* 12 (2): 295-310.
9. Morgan, P.R. 1972. 'Causes of mortality in the endemic Tilapias of Lake Chilwa.' *Hydrobiologia* 40 (1): 101-19.
10. Ruttner, F. 1931. 'Hydrographische und Hydrochemische Beobachtungen auf Java, Sumatra, und Bali.' *Arch. fur Hydriobiol.* Supplement Bd VIII.

Chapter 5 LAKES (II)

Lake Victoria is the greatest of the African lakes, and its level is now stabilised by the Owen Falls Dam. The average depth of the lake is only eighty metres in an area of 67,000 $Km.^2$; but despite this, there is some stratification in the northern part of the lake. Though the lake, as a whole, is not very fertile, there are bays and inlets, and especially the Kavirondo Gulf at the northeast of the lake, where fish production is high. For many decades, the indigenous species, the 'lake fish' *par excellence, Tilapia esculenta* and *T. variabilis*, have been top grade fish acceptable to all.

Lake Victoria Fisheries

Before the coming of the Europeans and the development of the country, the lake fisheries would have been on a subsistence basis, the riparian tribes catching fish for their own needs, and probably bartering a surplus. Their method of smoking *Tilapia* shows long usage. But then came the railway, and the establishment of towns, where a high spending market for fresh fish, and improved communications, encouraged a fish trade, organised by Asian middlemen. Subsistence fishing had now become an industry. Very high rates of fish capture were reported from these early days of the industry; for example, 100 fish per net in the Kavirondo Gulf. There is no need to doubt these figures, for catch rates as high as these can be seen today in little-fished African lakes. But from then onward, the *Tilapia* fishery of Lake Victoria followed the fate of all heavily-exploited fisheries in which there is no individual ownership of the resource.

By about 1928, the catch per net had fallen to about nine fish, in part because the number of nets in use was increasing. Attempts were made to enforce the use of nets with a large mesh, to prevent the wasteful capture of the fish while still small. Then came, first, the economic depression years of the early 1930s, when trade languished and fewer nets were in use; then came the war years, when nets were hard to procure. Because the number of nets in use declined during these years, the catch per net rose to a much more remunerative level.[1] Had it then been possible to licence and limit the number of nets, the *Tilapia* fishery could have been stabilised at a remunerative level indefinitely. But in a fishery like this, which began, and must have continued, as a subsistence fishery and a traditional right of the lakeside dwellers, any kind of regulation would be resented and evaded. Attempts to regulate the size of mesh allowable, and above all to stop a form of fishing in which many nets were joined end to end and dragged to the shore like a seine net, always failed, in spite of many prosecutions. Any such regulation would have to be explained patiently to the fishermen, as the means of improving the productivity and profitability of the fishery. It could not be enforced without the willing cooperation of the fishermen themselves. At present, the catch of *Tilapia* per net has fallen to much less than one fish. Yet, as the recovery of many fisheries has shown, so long as any *Tilapia* survive, the fishery will recover if the fishing effort is restricted and thereafter controlled, by the consent and cooperation of the fishermen.

Lake Victoria is also the home of a 'species flock' of small cichlid fish belonging to the genus *Haplochromis* and allied gendera. Almost all of these are endemic to this lake, and they show a wide range of feeding habits, thus making use of the different kinds of production of the lake. But they are all small fish, though relished by the Africans, who smoke them and sell them mounted on frames of twigs. In addition, there are large resources of a small silvery member of the carp family, *Engraulicypris*, which has adapted to a plankton feeding role; these resources have been surveyed by echo sounder. A trawling industry is being developed in the lake, chiefly in the deeper water, to exploit these fish and others larger. But of course this fishery will be mechanised, will employ few fishermen and do little to provide alternative employment for the inshore fishermen, so easing the pressure on the *Tilapia* stocks (Fig. 10).

A recent feature of the lake has been the appearance of the Nile perch, *Lates niloticus*, a fine sporting fish which inhabited Lake Victoria in an earlier geological period but was barred from the modern lake by the Ripon and Owen Falls, which prevented its ascent from Lake Kioga lower down the Nile. It has often been suggested that this predatory fish of high market and sporting value should be introduced into Lake Victoria, partly to crop down the big resources of small fish and thus turn them indirectly into valuable fish flesh; and partly to offer an important tourist and sporting amenity. Though for long years this proposed introduction was fought successfully, on the sound grounds that the depredations of such a formidable predatory fish could not be accurately foreseen, the fish finally appeared in Lake Victoria and spread rapidly there. This could have been the result of clandestine stocking, or by the fish ascending the turbines at the Owen Falls Dam. It appears that, at certain stages of the operation of the turbines, a fast swimmer could get through; in fact, at least one other fish, of zoological interest only, has done so. At all events, the Nile perch is now well-established in Lake Victoria and is an important commercial fish there. It is also a sporting attraction. It was feared that it would feed on great numbers of young *Tilapia*, rather than on the big shoals of *Haplochromis*; but it seems that the predator takes any small fish without discrimination.

Lake Tanganyika Fisheries

In Lake Tanganyika, on the other hand, the important fishery is for pelagic fish, as indeed one would expect, since the lake bottom deeper than about 200 metres is without oxygen and cannot support any kind of fish life. There are two species of freshwater clupeoids, genuine freshwater sprats, and four species of predatory fish allied to the Nile perch but endemic to Lake Tanganyika. There has been a recent study of the fisheries for these species in the southern, Zambian, part of the lake.[2] The two freshwater sprats are small fish never exceeding about 15 cm. in length and mostly much smaller. They have a short life cycle, accomplished in one year in most cases, and closely related to the periodic and spatial variations in the production of the phytoplankton and zooplankton on which they feed. The smaller of the two species, *Stolothrissa*, can feed on planktonic bacteria as well as on the microplankton. The larger species is more of a zooplankton feeder and predator, eating small *Stolothrissa* as well as other prey. But, in general, the two species are competitive for their planktonic food, and can replace one another; as they have the same commercial value, the fishery is not affected. The average catch of these two pelagic species is about 80 kg./ha./year, a high figure for a natural fishery.

The primitive fishery for these clupeoid fish was at night. The dugout canoes used had a firebox mounted on a kind of bowsprit, and in this a fire of dry and smokeless wood was kept burning. This light attracted the fish, just as, in the sea, herrings and other clupeoids are attracted by light; and when enough fish had gathered under the light, they would be scooped out with a long-handled dipnet. On a breezy night, sparks from the furnace go flying downwind over the canoe and its crew, and burn holes in the net. A great improvement on this was the introduction of paraffin pressure lamps by European fishery officers; this was not only a much more efficient attractor of fish, but a relief to the Forestry Officers, who were becoming concerned by the felling of this special timber for the flares; forty headloads might be used per village per night.

Industrial fishing for these clupeoids began in the then Belgian Congo where Greek fishermen introduced the purse seine used from sizeable diesel launches. The purse seine is a big net which can be set around a school of fish attracted by lights, and then pursed to enclose and catch the fish. Catches were at the rate of tons per shot, but there were no signs that this exploitation was adversely affecting the stocks of fish (Fig. 10).

Indeed, with such a short food chain and such a fast natural turnover, heavy fishing could be sustained; for the same reason, marine clupeoids stand much heavier fishing than bottom-dwelling fish.

The use of the purse seine began in the Zambian waters of the lake in about 1962, and a form of encircling net, the *chilimila* net, was also introduced from Lake Malawi. The dipnet remained in full use, especially in the coastal waters of the lake. The biomass of the clupeoids reach an annual maximum in September, and *Stolothrissa* was the dominant fish in all but two of the years of sampling. The four species of perchlike predators of the clupeoids decreased in numbers from 1963 to 1966, and have remained low since. But the catch of the clupeoids increased from 1964 to 1967, and has remained high since. The observed reduction in the fish of prey was ascribed to a reduced rate of survival under the increased fishing pressure, with a corresponding increase in the abundance of the clupeoids, their prey. The commercial fishery in the Zambian waters of Lake Tanganyika is in a stable phase, in which the average nightly catch of all species varies little from year to year, the biomass of the clupeoids being much greater, and that of the predators less, than in the pre-exploitation fishery. This is a much more agreeable picture than that of the Lake Victoria *Tilapia* fishery.

Luckily, as yet no great industries have established themselves to pollute these deep African lakes; for remedies would be almost impossible. Very different has been the fate of the great American lakes, such as Lake Michigan, which is now in the area of one of the most heavily industrialised regions in the world.

The American Great Lakes

In the early 1800s, Lake Michigan supported a substantial population of salmonid fish, so that a large and well-oxygenated hypolimnion must then have existed. Since then, these large stocks of fish, which bred in the affluent streams, have largely disappeared, and the present fish population is much affected by pollution, including eutrophication by pollutants, by intensive commercial and sport fishing, and by the introduction of non-native species of fish.[3]

The five great Laurentian lakes of North America, Lakes Ontario, Erie, Huron, Michigan and Superior, are together by far the largest body of fresh water in the world, covering about 245,000 $km.^2$ Originally, only Lake Ontario had a direct connection with the sea by way of the St Lawrence River, and therefore was the only lake with natural runs of salmon and sturgeon, and immigration from the sea of alewives, sea lampreys, smelts, elvers and other marine and brackish water fish. At that time, also, the human population of the drainage basin of the Grest Lakes was sparse and their effect on the environment negligible.

Aside from the marine component of Lake Ontario, the principal species of the original fish fauna of the Great Lakes in about the year 1800 were the lake whitefish, *Coregonus clupeiformis*, lake herring, *C. artedi*, lake trout, *Salvelinus namaycush*, and cottids or bullheads, ciscoes or chubs (*Coregonus* spp), and *Prosopium* spp, round or pygmy whitefish, of small direct commercial value, but important as forage fish for the big predators. Other less prized fish were the burbot, pike and some small cyprinid fish which, like the ciscoes above, were chiefly important as forage fish.

But this natural environment was soon upset by the restless activity of the early settlers. The felling of the forests and the clearing of the land for agriculture much reduced the rate of flow of the rivers and streams, and raised their temperature in the summer by as much as 8°C. Milldams and the unrestricted dumping of sawdust and other wastes changed the character of the streams as breeding places for the lake dwelling fish. Many of the well known spawning runs of lake sturgeon, and of the big and valuable coregonid and salmonid fish, were stopped, and the high temperatures of the streams now inhibited the spawning of many other species. The deposits of silt and sawdust put an end to the river runs of lake trout, whitefish and lake herring by the early 1900s. The earliest commercial and subsistence fisheries were by weir, dragnet and hand nets.

Undesirable Immigrants

In the early 1800s, canal building was the current technology, as it also was in England; and the Welland Canal was dug to enable shipping to pass the Niagara Falls into the upper lakes. At about the same time, a canal to link the headwaters of the Hudson River with Lakes Ontario and Erie was dug. These canals also gave access to the lakes by many fish of marine origin. The most notorious and destructive was the sea lamprey, which flourished and became a dangerous parasite of the commercially valuable lake fish; for instance, it reduced the population of the lake trout and lake whitefish to very low levels.

The spawning of the sea lamprey was favoured by the now warmer waters of the streams, the same factor which made those streams unsuitable for the spawning of the native lake fish. The alewife, *Alosa pseudoharengus*, a marine member of the shad family and a plankton feeder, ascended the rivers from the Atlantic in spring to spawn in ponds and sluggish rivers of the eastern north Atlantic from Newfoundland to Florida, but spent most of its life in the sea. It is an important food fish; but about the time of the building of the canals it established a landlocked population in the Great Lakes, which competed all too successfully with the native lake plankton feeders, such as the lake herring and the deep water ciscoes.

Yet another estuarine immigrant was the smelt, *Osmerus mordax*, which feeds principally on the larger plankton animals.

The Effect of Commercial Fishing

When commercial fishing in the Great Lakes was established, the fish population was drastically affected, for such fishing is not indiscriminate, but concentrates on the capture of the most popular and valuable fish. The lake sturgeon was also intensively hunted, but for a different reason; it was such a nuisance, damaging nets laid for better fish, that it was destroyed when caught. After the salmon, the lake whitefish was the most valuable species, and was fished in place of salmon as this became rarer; next in value was the lake trout. The ciscoes were also heavily exploited. Still later, when intensive fishing had reduced the stocks of all the more valuable fish, perch and the pikes, though never regarded as prime fish, were heavily fished.

The species which has best withstood the impact of heavy commercial fishing in the Great Lakes is the lake trout, a fish at the apex of the food pyramid which feeds, when young, chiefly on the very abundant large crustacea, *Mysis relicta* and *Pontoporeia affinis*. This food chain can be carried one stage further back; for it has just been shown that *Mysis relicta* winnows out and feeds on organic material from sediment.[4] When well grown lake trout feed on deep water ciscoes (*Coregonus* spp) and deep water sculpins (*Myxocephalus* spp).

When the native predatory fish, the lake whitefish, lake herring and, to a lesser extent, the lake trout, were reduced in numbers by the depredations of the sea lampreys, the immigrant alewives, freed from predation, became extremely abundant, and a considerable nuisance and a health and recreation hazard. The shoals choke water pipes and foul the beaches with mass kills. To crop them down and thence to provide first class food and sport, Pacific salmon species have been introduced. The State of Michigan alone introduced in 1970 some 500,000 small coho and 170,000 chinook salmon, as well as 275,000 steelhead trout (*Salmo gairdneri*). The salmon have not bred, but their numbers are maintained by large annual plantings of young. In Lake Superior, the pink salmon has been introduced and is breeding. It is not a popular fish, and will inevitably compete for food with the native lake herring.

Eutrophication

It is not understood how eutrophication can bring about the decline or extinction of fish in an oligotrophic lake; but the decline has followed the same course in all the American Great Lakes. The first eutrophication of the lakes followed on the intensive settlement of the land, and led to the extinction of the migrant Atlantic salmon. All the Great Lakes now show an increase in their content of phosphorus and nitrogen, and a decrease in the dissolved oxygen content of their hypolimnion.

Lake Baikal

That other great oligotrophic lake, Baikal, in eastern Siberia, the deepest and one of the largest in the world, lies near the cold pole of the northern hemisphere and is frozen over for many months of each year. It has many affluent rivers, of which only one is of major volume, and it has a major river, the Angara, as its outlet. The Angara is now dammed for the Bratsk hydroelectric scheme, which is claimed to be the world's greatest, and Lake Baikal itself is now the reservoir for this station, just as Lake Victoria is now the reservoir for the Owen Falls hydroelectric scheme.

Baikal is a very old lake, in which an endemic fauna and flora has evolved. Out of a total of about 600 plant species living in the lake, some 200 are exclusive to Baikal; and of 1,200 animal species, some 750 are endemics.[5]

Most of the lake is very deep, for 64 per cent of the lake area is more than 500 metres deep, and 25 per cent more than 1,000 metres deep. Only eight per cent of the lake area is less than 50 metres deep. However, the few bays of shallow water play a disproportionately large part in the productivity of the lake. Whereas the open waters do not increase in temperature in summer above 4° to 5°C, these bays warm up to 18° to 20°C, and this warmed water spreads out over the open lake. In this shallow water live the familiar freshwater fish fauna of burbot, pike, roach, perch and grayling.

The water of the lake is pure, containing only 100 gm./litre of dissolved salts and a low content of organic matter. The water is so clear that photosynthesis takes place at 60 to 70 metres depth, though naturally it is greatest at 20 to 25 metres. The primary production of the lake has been estimated at 122.5 gm.C/m.^{2} /year; in such a large lake, this is equivalent to a production of four million tons of carbon or nearly eight million tons of algal wet weight. This high value is due to the great depth of the productive layer and the turbulent mixing of the lake water which is always bringing up new nutrients to keep production going. There is only one annual outburst of phytoplankton growth, in contrast to two, in spring and autumn respectively, in temperate lakes. The main phytoplankton outgrowth is in April-May, and is due to the diatom *Melosira*, which grows beneath the ice cover and has its maximum in May. This great volume of 'net plankton' is supplemented by a summer growth of nanoplankton which exceeds by four to six times the biomass of the spring net plankton. This nanoplankton, and also a bacterioplankton, maintains a bountiful supply of food for the filter feeding copepods until well into the autumn.

The bacterioplankton swarm in summer to feed on the decaying net plankton, and in June and August they may exceed 500,000 cells/ml. The activity of these bacteria, and the high content of the water in dissolved oxygen, result in very little organic matter reaching the bottom mud of the lake.

The next step up the simple food chain of the lake is the filter feeding copepod *Epischura baicalensis*, which feeds both on the net plankton, and on the nanoplankton and the bacterioplankton. By September, the average biomass of zooplankton reaches 23.9 gm./m.^{2} and when the copepods make their nightly migration to the surface they may number 30,000 to 50,000 individuals per cubic metre, a density equal to that found in some eutrophic lakes.

The next most important members of the zooplankton are the predatory amphipod *Macrohectopus* and the copepod *Cyclops*. The amphipod consumes about 600,000 tons of zooplankton annually in Lake Baikal.

The principal fish are cottoids, species of *Cottomephorus* and *Comephorus*, all species endemic to Lake Baikal. All are feeders on zooplankton, and because they do not swim in shoals, they are not fished for commercially. They do, however, play an important part as the principal food of the Baikal seal, *Pusa sibirica*, of which the herd in the lake averages about 33,000 individuals. This herd is culled to give about 2,100 carcasses annually. The seals take as food about 33,000 tons of fish, that is, each seal eats about one ton of fish per year, and of this about 22,000 tons are *Comephorus* and 8,000 are pelagic cottids.

The most important commercial fish in the lake is the celebrated omul, *Coregonus autumnalis*, a notable delicacy. This fish is a plankton feeder, which ascends to the surface waters in summer and autumn, when the zooplankton is at its optimum density; but descends to spend the rest of the year at 100 to 300 metres depth. The average biomass of omul in the lake is about 25,000 tons, but desultory fishing results in little of this stock being caught. To build up these stocks further, all omul fishing was prohibited until 1969. Artificial breeding may be resorted to; and there may be transplantation of landlocked species of salmon, as in the American Great Lakes. By such means, it is hoped to increase the harvest of salmonid fish to 5 to 6 kg./ha. annually, as compared with 1.6 kg./ha. for the production of omul.

There has been some recent concern about the future of Lake Baikal.[6] The wildlife in the lake basin is threatened by industrialisation of the lake shores, and by the heavy felling of the forests, which, again as in North America, has resulted in the drying up of many of the affluent streams. It is now proposed to make Lake Baikal a national park.

NOTES

1. *Annual Reports of Game and Fisheries*, Uganda.
2. Coulter, G.E. 1970. 'Population changes within a group of fish species in Lake Tanganyika.' *J. Fish. Biol.* 2: 329-53.
3. Smith, S.H. 1972. 'Factors of Ecologic succession in oligotrophic fish communities of the Laurentian Great Lakes.' *J. Fish. Res. Bd. of Canade.* 29: 717-30.
4. Lasenby, D.C. and Langford, R.R. 1973. 'Feeding and assimilation in *Mysis relicta*.' *Limn. Ocean.* 18 (2): 280-5
5. Moskalenko, B. and Votinsev, K. 1970. 'Biological productivity and balance of organic substance and energy in Lake Baikal. Productivity problems of Fresh waters.' *IBP/UNESCO Symposium*. Dalny, Poland. May, 1970: 207-25.
6. Anon. 1969. 'The Future of Lake Baikal.' *Nature* 223: 1091.

Chapter 6 ARTIFICIAL DAMS AND LAKES (I)

The natural supply of fresh water is usually variable, with abundant or even excessive quantities at one season, and periods of drought in another. Moreover, there are also longterm variations, in which water may be deficient for more than one season. But for all purposes of human settlement and industry, water is needed at a steady rate; so, for millennia, men have sought to store water in times of abundance (thereby also controlling destructive floods) in order to draw upon it in times of scarcity. Many of these works are very ancient. Lake Qarun in Egypt (p. 94) and Lake Habbaniyeh in Mesopotamia, were both used in ancient times as flood control reservoirs, by diverting water into them at high flood, so that, as the flood levels fell, the stored water would discharge into the river and so even out its flow.

In India and Ceylon, also, great water storage works were built, to hold water especially for irrigation of crops in the dry season, thus assuring year-round cultivation of the land. The monarchs of those distant days recognised that an assured source of food was also a source of power. It was also the source of the first civilisations. For as soon as the farmer could produce more food than he required himself, he set free others to develop culture and letters, and the management of the irrigation itself developed mathematics.

Irrigation schemes are still regarded as one of a country's best investments. In modern times, great barrages are built across rivers, creating artificial lakes which even out the river flow, divert water into irrigation systems, maintain a depth of water for navigation and flush out pollutants and silt.

Reservoirs are either for a special purpose, or are multipurpose. Drinking water reservoirs used to be kept for that purpose only; but the trend nowadays is to open all reservoirs to recreation and fishing, including those built for the storage of domestic and industrial water.

In the United States especially the use of reservoirs for recreation and leisure has become big business, since most recreation is water based. Wherever possible, they have been stocked with fish to provide sport, and, to a much lesser degree, some commercial fisheries. The results are impressive. In warm water reservoirs, the average rate of fish production was 15.7 lb./acre (14 kg./ha.) in 1960, and in cold water reservoirs, stocked chiefly with salmonids, 9.0 lb./acre (8 kg./ha).

Flood control reservoirs are not well suited for fisheries, for by definition they should be empty, or nearly so, to hold the maximum flood waters when these come. They must then have a great rise and fall of level, and at low water their banks are unsightly deserts, since neither land nor water vegetation has a chance to establish itself. However, flood control reservoirs on the Columbia River, in the USA, have been successfully used for the rearing of young chinook and coho salmon, adapting their management to the management of the reservoir. The survival of the young to return from their sojourn in the ocean was nearly as good as that of expensive hatchery-reared young salmon.[1]

The most important reservoirs are those built for the purposes of irrigation and hydroelectric power generation.

Reservoirs for Irrigation

These are more usual in lowland areas, close to the flat lands to be irrigated. They must have a large annual drawdown as the water is drawn off for use, and they are often of very large area. Such were the old tanks in Ceylon, where one was called the 'Sea of Pakrasi.' Their productive use as fisheries depends on the fact that, even at drawdown, some water is left, in which enough fish survive to breed as soon as new water flows in; and, in a warm climate and with rich feeding, the offspring are of edible size by the next drawdown.

Fishing is discouraged in the canals, for fear of blockage and siltation if the current is slowed down; however liftnets are used, and prawn traps.

Fish in Irrigation Works

Fish and prawns abound in irrigation works, reservoirs, channels and even in the irrigated fields while under water. In Malaysia and Indonesia, for example, the fish which grow in the warm and well-manured shallow waters of ricefields are a valuable crop, in some cases as valuable as the rice. There are cases where the landlord can take the rice crop if he will agree to let the tenant keep the crop of fish. There may be an artificial stocking of ricefields with fish such as the common carp; it is even maintained that these fish, by rooting about among the growing rice plants for insects, not only help to control these insects, but may stir up the soil and so release nutrients beneficial to the rice.[2] In Malaysia, early in the present century, the air-breathing gourami, *Trichogaster pectoralis*, the Sepat Siam, was introduced, as its name implies, from Siam or Thailand, because it grows much larger than the native Sepat padi, or *Trichogaster trichopterus*. Naturally, in the environment of a paddy field, swamp conditions prevail, with oxygen in short supply; so paddy field fishes include those with accessory air breathing organs to supplement the gills. These include not only the small goramies mentioned above, but predators such as the snakeheads and catfish, the former of which fetches a good price on the markets. Cast nets and lift nets in the irrigation canals take a good fish harvest, and the paddy fields themselves are arranged so that there is a deeper sump into which the fish accumulate as the fields are dried out for harvest. Then the contents of the sump, a greasy mixture of mud and fish, are baled out over a sieve, and the fish sold on the spot to dealers who will smoke them for the Indonesian market.

Dams for Power Generation

For power generation, which also needs the conversion of an intermittent water supply into a steady one, a higher head of water is needed than for an irrigation reservoir; and so streams and rivers are dammed to give the necessary head. Everyone has seen water mills, now in disuse in Britain, where a weir across a stream raises its level, and a diversion channel, the mill leat, leads the water to the mill wheel which generates the power. But the demands for power are very great and increasing, so the water mill is now superseded by the power station, and the mill wheel by turbines. Rivers are dammed with massive barriers, behind which artificial lakes, of areas comparable with the great natural lakes of the world, build up. Many such dams can be used both for power generation and for irrigation. The Aswan High Dam, which has created the artificial Lake Nasser, is the most notable case.

But while the storage reservoir of this huge project will allow of the cultivation of hundreds of thousands of new hectares, some of the power will have to be used to manufacture fertilisers. This is because the fertilising silt carried by the Nile flood will now be deposited mostly in the reservoir, and no longer spread its wealth over the fields, as it has for past millennia.

These artificial lakes, built for the generation of electricity, are designed accordingly. The planned residence of time for the water in the lake will depend on the calculations of the engineers, who will balance the predicted water supply against the requirements of the power station, with enough water in hand to give the head required and to guard against unnatural drought periods. Obviously, they will plan for the minimum safe residence time for the water in the lake; for an excessive amount would put up the capital cost of the scheme. Thus, the Kainji Dam on the River Niger is planned to give the water a residence period of only seventy days; in the Kariba Dam it is three years.

Many dams are planned for multiple use, especially in America. The great Roseires Dam on the Blue Nile had provision for a power station and the Ord Dam, in northwestern Australia, though at present for irrigation only has provision for a hydroelectric power station when the demand for power justifies the building of one.

But these artificial lakes are also of great interest and importance as fisheries, and the residence time of the water affects the performance of the lake as a fish provider. In natural lakes, as shown in the last Chapter, the residence time may be measured in centuries, and the water takes on a special character, giving an environment in such settled conditions different from that of flowing waters and usually more productive. In dealing with the difficulties and problems of artificial lakes as biological units, it has to be remembered that they were designed to give the optimum power, not the optimum fish.

Interference with Natural Flood Regimes

Artificial lakes differ from natural lakes in that the deepest point is at the foot of the dam wall, and not somewhere in mid-lake; that the new lake drowns large or very large areas of land, not seasonally, but permanently; and that in most of them the fauna and especially the fish fauna, is in a state of rapid change, adjusting from a flowing to a stable environment. In Chapter 3 it was shown that the seasonal flooding of the land in the river valleys was an essential element in their fertility, and also contributed to some of the greatest freshwater fisheries in the world. In these artificial lakes, the annual 'drawn-down' may expose seasonally a relatively small area of the former valley, though this exposure has important effects on the fish production. Finally, by stabilising and evening out natural fluctuations in the discharge of rivers, dams may put an end to seasonal natural irrigation and to the flood fisheries below the dam.

Social Resettlement Problems

The clearing of the land which will be flooded and the resettlement of the displaced population pose considerable social problems, and involve long-term plans for compensation and alternative living space. There have been complaints that in no case yet has enough time been given to the social aspect of dam construction.[3] In Lake Kariba, for instance, there was violent opposition to the taking of the land in the Gwembe Valley for the new lake, in spite of studies to find fertile pockets of land for resettlement and generous compensation. Some conservationists consider that this dam was a mistake, and that the loss of animal protein by the elimination of the game in the valley (aside from those 'saved' by the hopeless but well-publicised 'Operation Noah') will be greater than the gain due to fish production in the new lake. But food was not the main consideration; and conservationists who deplore damage to the environment should remember that men are also a part of the environment, and that the mass media and easy communications have led to expectations of rising standards of living (however these may be defined) which will not be foregone. A rising standard of living is associated with plentiful power, which spreads employment as well as comfort and convenience.

The great dams built for power and irrigation in Siberia will have displaced few, for these are mostly in sparsely inhabited parts of the country; but the damming of the River Volta, to generate electricity for the State of Ghana and for an important aluminium smelter displaced 70,000 people; the Aswan High Dam displaced 120,000.

The Zambesi River and Kariba Dam

The Zambesi River in the Gwembe Valley has been described as a 'sand-bank river,' with well-defined steep banks, and a bed composed largely of sandbanks, which, at low water, would break up into a series of large lagoons connected by the much reduced river flow.[4] The river had a great seasonal amplitude of water level and at high water a considerable area of the valley was flooded. The rise of the river at the beginning of the floods was the trigger for spawning for the most important species of fish, which was of short duration. In the floods, the young newly-spawned fish found abundant food, and also space and shelter from the many predatory fish. As the floods receded, the fish returned to the main river and some were caught in primitive weirs made of boughs and grass set in the depressions in the ground down which the water found its way back to the river. Similar and larger traps were set in the falling estuaries. And, of course, the villagers planted the banks, newly-irrigated and with a fertilising deposit of river silt,with their cereal crops.

Predator-Prey Alternation

At low water, all the fish were gathered together in the lagoons, where the predatory fish had their own way, for their prey had no escape and no cover.[5] In consequence, by the end of the dry season, a reduced fish population included a high proportion of the predators. A similar regime was described for the flood fisheries in Borneo in Chapter 3. There was in effect an alternation in the fish populations, between a dry season population with a high proportion of predators, of which the tiger fish, *Hydrocynus*, was the most voracious; and a wet season during which the nonpredators recovered.[6] Because of this predation, the average size and weight of the fish was comparatively high. In some ex-preimental fishings with a fleet of gillnets in two flooded backwaters of the pre-impoundment Zambesi, 303 fish were caught with little trouble in 1956; although these were caught in the height of the flood season, 87 of these fish, or 28 per cent, were the predatory tiger fish and catfish. The average weight of the tiger fish was 510 gm. and of the nonpredators *Labeo* and *Citharinus* 724 and 510 gm. respectively – a large run of fish. The dam was closed in December 1958, and thenceforward there were no more low water periods but always a rising water level.

During a second visit to the lake in May, 1960, not two years after the dam closure, fishing gave different proportions of these fish. The predators were now 36 per cent of the catch, and the average weight of the tiger fish was 820 gm. But now *Tilapia*, of which only two specimens were caught among 303 fish in 1956, was equal in numbers and average weight to the tiger fish, at over 800 gm. each. Commercial fishing had just begun, and the fish being sold and smoked were chiefly tiger fish and *Tilapia*. In 1964 and 1965, the nonpredatory vegetarian species *Tilapia* and *Labeo* comprised over 70 per cent by weight of the commercial catches of fish. This eclipse of the predators was due to a better rate of survival of the young of their prey, in a general increase in the abundance of all the former river fish species.

The Salvinia Pest

Thus, the fishery in the new lake got off to a good start. But meanwhile an unwelcome development was the appearance and very rapid spread of the water fern *Salvinia*, which soon covered 200 square miles of the lake's surface, and had an adverse effect on the fisheries, since the weed beds clogged the nets.[7] The general alarm caused by the weed led to the spending of large sums on contractors to control the weed. In fact, the weed spontaneously ceased to spread and is now most persistent where it has accumulated among the branches of the drowned trees. Between these fixed mats of weed, there is a clear zone close to the shore where fishing with gill nets is possible. On the weed mats, other plants such as grasses and shrubs are growing and binding the mats into a 'sudd.' But the volume of the weed in the whole lake is very great; and it has been shown that *Salvinia* is a 'luxury consumer' of nitrogen and phosphorus nutrients, accumulating far larger quantities than they require; they have been described as 'nutrient sumps.' Perhaps because of this loss of nutrients, the fisheries of the lake have been taking fewer fish since the earlier years. The catches on the northern (Zambian) shore reached their peak of 3,700 tons in 1963, but have since then fallen to 900 tons in 1967, with a recovery to 2,063 tons in 1970 and 1,500 in 1971. There are no more recent figures. Of course, some part of this decline will have been the result of declining catches which discouraged fishermen and reduced the fishing effort.

Instead of increasing their fishing effort to make up for falling catches, many of the fishermen have become part-time. But a nucleus of professional fishermen is emerging.

A contrary view is that the nutrients locked up in the *Salvinia* mats have been saved from being washed out of the lake through the turbines, and may yet play an important part as the weed beds rot down. In the meantime, the weed beds, in spite of their complete cover, and the impossibility of photosynthesis taking place in the water beneath them, still allow of a 37 per cent saturation with dissolved oxygen two miles into a weed mat,[8] which would allow many river fish to survive; and it has been found that the weeds are a substrate for a flora and fauna of epiphytes and invertebrates which contribute to the productivity of the lake. As will be shown, other man-made lakes also showed a decline in fish production from a first flush season, though they have not as yet been afflicted by plant invasions on the scale of the *Salvinia* on Lake Kariba.

Changes in the Composition of the New Lake Waters

In the River Zambesi before the closing of the dam, the water was rather poor biologically, with a low content of dissolved solids at twenty-six parts per million. In the river at low water the content of the water was higher at sixty-seven p.p.m., and when the dam was closed the impounded water at first had a content of sixty-five p.p.m.[9] It rose to sixty-nine in 1958; but thereafter declined year by year to forty-two in 1964. No doubt the initial high values were due to the leaching out from the newly-flooded soil of the valley of the nutrients and wood ash resulting from the burning of the trees felled to make clear fishing pitches. The outburst of *Salvinia* took advantage of this comparatively high level of dissolved nutrients, and must have done much to reduce it. During the first years, also, the waters of the lake stratified, and in the hottest months a well-marked thermocline formed at about ten metres, below which the hypolimnion was totally deoxygenated. No doubt the rotting of large quantities of vegetation and organic matter contributed to this deoxygenation. But, when the cooler months of the year came round, the surface layer cooled until there was a complete overturn of the lake waters, so that some dissolved oxygen was distributed over the whole lake and to the lake bottom. The thermocline has become a regular feature, but as the years have passed, the period of deoxygenation of the hypolimnion has become progressively shorter, and the bottom water of the lake has become accessible to the fish population. This distribution of the fish over a wider area, and to depths not accessible to the gill nets of the fishermen may have been partly responsible for the observed fall in the catches in the lake. It is a fact that echo sounder traces show fish in water deeper than twenty metres, which would be about the limit for surface set gill nets.

Productivity of Lake Kariba

The productivity of the lake is due, not only to the phytoplankton, but to the rotting trees and to the mud which is forming. Both these are helped by the annual draw-down, a variation in lake level of about ten metres due to the different rates of flood water entering the lake and of the discharge of water through the turbines and spillways.

The lake bottom is still heavily obstructed by drowned trees, which are clearly to be seen in echo sounder tracings of the bottom of the lake. These trees may take centuries to decay completely. I have seen the Chenderoh Dam in Malaysia, after sixty years of flooding; the trees in the dam are riddled with insect holes but are still standing. In Lake Kariba and the other lakes these rotting trees are likely to make a great contribution to production. Their leaves will long since have decayed to detritus which will help to produce an organic mud on the bottom; but their bare branches will form a substrate of very great total area for the growth of aufwuchs, the living felt of epiphytic plants and their associated animals, where there is light enough for plant growth. In fact, by 1966 this aufwuchs was already growing well on the submerged branches.[10] But the wood itself harbours animal life, and especially the larvae of chironomid midges, to the extent of 97 mg./m.^{2}, after only four months. This is a modest rate of production per unit area; but from millions of branches spread over the lake slopes the aggregate production will be very important.

This production is enhanced by the draw-down, which changes the fauna of the rotting trees. Wood boring insects, and above all the Ephemeropteran *Povilla adusta*, at reflooding produce a biomass of 76 mg./m.^{3} whereas the bark produced a fauna of insects and worms with a biomass of 22 mg./m^3.

The Annual Draw-Down

The annual draw-down of about ten metres exposes about 300,000 acres of the lake bottom for some weeks. A quick growth of a good grass sward spreads over these areas, and attracts herds of game, including buffalo, elephant and antelope. The plentiful dung of these grazers produces a local fertility such that the catches of two of the lake's herbivores, *Tilapia* and *Haplochromis darlingi*, are many times higher off such game areas than elsewhere on the lake shores.[11] Further, in the shallow water chironomid midges are very plentiful. As the lake level rises at the end of the draw-down, the five or six small species of these larvae are replaced by one species of large larvae, resulting in a great increase of biomass of organisms very useful as fish food.

True Underwater Plants

As a true mud developed in the lake, underwater plants became established, which also provide a substrate for an aufwuchs. The submerged macrophytes include *Potamogeton, Ceratophyllum* and *Ludwigia*. The *Ludwigia* and of course the *Salvinia* already described, are floating plants or have floating leaves, while *Potamogeton* and *Ceratophyllum* are submerged. The establishment of these plants has much increased the biomass of the invertebrates of the lake.[12] Beneath the floating leaves was a fauna with a dry weight of 383 mg./100 gm. of plant; and under the submerged rooted plants it was 52 mg./100 gm. A mud fauna, chiefly of chironomid larvae, became established, and the submerged foliage was an equally good medium for animal life, but here fewer chironomid larvae and more snails and insect larvae. Under the *Salvinia* sudd, few invertebrates occurred; but in spite of this, areas associated with the weed had a high biomass per unit area, up to five times as high as in areas where *Salvinia* is absent. It would almost seem as if the weed, so alarming at first, may be settling down as a modest benefactor.

Fish Species

Lake Kariba had twenty species of fish when first filled, but only nine of these have had any commercial importance. No plankton feeder was in this group, so an attempt was made to introduce another *Tilapia, T. macrocheir*, which was believed to have this habit; and *T. melanopleura*, which is a herbivore. In spite of the introduction of thousands of pond reared young, neither of these introductions was successful. *T. mossambica* takes planktonic algae as well as epiphytics, and, as stated earlier, underwent a massive increase in numbers. Meanwhile, the native *Alestes imberi* was ousted by the small and prolific plankton eating *A. lateralis*, a migrant from the Upper Zambesi.[13] It was already establishing itself as the plankton eater of the open lake, when the freshwater sprats of Lake Tanganyika, mentioned in Chapter 5, were successfully transplanted to Lake Kariba, soon becoming abundant. The Kariba fishermen are being taught to catch them with the aid of bright lights to attract them for capture, as on Lake Tanganyika;[14] and a fishing gear from Lake Malawi, the *chilimila*, also mentioned in Chapter 5. These sardines have already become the principal food of the tiger fish and may cause a second increase in the populations of these valuable fish. They may also populate the waters of the open lake and make use of the phytoplankton and zooplankton growing there. Freshwater prawns were also transplanted from Lake Tanganyika, but their fate is not yet known. The Russians have successfully transplanted useful fish food organisms into their artificial lakes, and the barren Canadian lakes have been made a better nursery for young salmon and trout by the transplantation of the crustaceans *Mysis relicta* and *Pontoporeia affinis*. Prawns in Lake Kariba would be able to feed on detritus, and be a good food for many species of fish. They might also be a saleable food for humans; most prawn species fetch high prices.

NOTES

1. Korn, L. and Smith, E. 1972. 'Rearing juvenile salmon in Columbia Basin storage reservoirs.' in *Reservoir Fisheries and Limnology*. Ed. Gordon E. Hall. Sp. Pubn. No. 8, Amer. Fish. Soc. Washington, D.C.: 287-98.
2. Hickling, C.F. 1971. *Fish Culture*. London. Faber & Faber, 1971: p. 226.
3. Balon, E.K. 1972. Illustrated resume of Technical exhibit at 102nd Annual Meeting of the American Fisheries Society. Hot Springs. Ark. Sept 10-13, 1972.
4. Jackson, P.B.N. 1960. 'Summary of Zambesi Valley surveys.' *J.F.R.O. Ann. Rept.* 8 (1958): 24-7. Govt. Printer Lusaka.
5. Jackson, P.B.N. 1960. 'The ecological effects of flooding by the Kariba Dam upon the middle Zambesi fishes.' *Proc 1st Fed. Sci. Congr.* Salisbury, Rhodesia: 277-84.
6. Jackson, P.B.N. 1960. 'Hydrobiological research at Kariba.' *New Scientist.* 7 (117): 877-80.
7. Gaudet, J. 1973. 'Growth of a floating aquatic weed *Salvinia* under standard conditions.' *Hydrobiologia* 41 (1): 77-106.
8. Harding, D. 1961. 'Limnological trends in Lake Kariba.' *Nature* 191: 119-21.
9. Harding, D. 1964. 'Hydrology and fisheries in Lake Kariba.' *Verh. int. Verein. Limnol.* 15: 139-49.
10. MacLachlan, A.J. 1970. 'Submerged trees as a substrate for benthic fauna in Lake Kariba.' *J. Applied Ecol.* 7: 253-66.
11. MacLachlan, A.J. 1970. 'Some effects of annual fluctuations of water-level on the larval chironomid communities of Lake Kariba.' *J. Animal. Ecol.* 39: 79-90.
12. MacLachlan, A.J. 1969. 'The Effect of aquatic macrophytes on the variety and abundance of the benthic fauna in Kariba.' *Arch. Hydrobiol.* 66 (2): 212-31.
13. Balon, E.K. 1971. 'Replacement of *Alestes imberi* Peters 1852 by *A. lateralis* Boulanger 1900, in Lake Kariba, with ecological notes.' *Fish. Res. Bull. Zambia.* 5: 119-62.
14. Balon, E.K. 1971. 'First catches of Lake Tanganyika clupeoids (Kapenta – *Limnothrissa miodon*) in Lake Kariba.' *Fish. Res. Bull. Zambia.* 5: 175-85.

Chapter 7 ARTIFICIAL DAMS AND LAKES (II)

Lake Volta

A dam across the River Volta, in Ghana, was suggested as long ago as 1916; but the programme was not completed until 1964, as a major hydroelectric station to supply electricity to an aluminium smelter and to the territory generally. More recently, channels are being cut to use water from the newly-formed Lake Volta in a large irrigation scheme in the Accra Plains. The dam was built at Akosombo, some sixty miles from the sea, and the lake it has dammed extends far up the valleys of the rivers Otu and White and Black Voltas. Lake Volta has an area of 8,840 $km.^2$, as against 5,364 $km.^2$ for Lake Kariba. It is also shallower with a mean depth of nineteen metres, compared with thirty metres in Lake Kariba. In Lake Volta, the water has a long resident period, and the upper parts of the lake keep their riverine character.

A thermocline formed very close to the surface soon after the lake filled, and the greater part of the volume of the lake was completely deoxygenated. But as the lake matured the deeper waters became oxygenated, and a rich bottom fauna, chiefly of the larvae of Chironomid midges, developed.[1] Aquatic vegetation began to grow in the more sheltered parts of the lake, and became associated with a varied and plentiful invertebrate fauna.

Productivity of Lake Volta

The lake had a plankton only in the surface waters so long as there was a thermocline. The phytoplankton consisted of either diatoms or blue-green algae, the two being a succession. Primary production became high. In 1970, it was measured at 2.0 to 5.2 $gm.C/m^2/day$, a figure which compares with the 2.0 to 4.0 of the fertile Lake Edward in Uganda.[2] But it is far surpassed by Lakes George and Albert in Uganda at 4.0 to 6.0, and by the White Nile at 10 $gm.C/m^2/day$.

Lake Volta is shallow and very near the Equator, so it will be warm and have a deep light penetration. This favourable geographical situation, the very long shoreline, the draw-down which exposes a wide area of the lake bottom to mineralisation and a long resident period of the water in the lake, all make for a high rate of production, which has shown itself by the rapid rise in the catches of fish in the lake to 50,000 tons. As with Kariba, this initial high level of productivity has since fallen and seems likely to stabilise at about 39,000 tons a year, a high figure enough, which has attracted thousands of fishermen to partake in such prosperous conditions.

As in Lake Kariba, the drowned trees have become an extensive substrate for the growth of aufwuchs and of a large insect population. Even before the dam was built, the two Volta Rivers were productive fish rivers, supporting large drag net and gill net fisheries. The principal species caught were the Citharinids, the Characins, and the predator Nile perch, *Lates niloticus*.

With the closing of the dam, and the beginning of lacustrine conditions, the fish fauna began to change. During the first two years, many of the riverine species, especially the Mormyrids and Characins, vanished and are now caught only where the rivers enter the lake. Most of the inhabitants of the rivers were either predators or insectivores. But in a very few years all this has changed; the dominant fish are now the herbivorous *Tilapias*, which comprise 50 per cent of the commercial catches.[3] The present dominant insectivorous fishes are clupeids, members of the herring family, and chiefly *Pellonula*, which are pelagic and surface feeding, small and delicate, but much prized as food fish. They feed on aerial insects as well as on the larvae of the Mayfly *Povilla adusta*, which grow profusely in the roots of the rotting trees of the former valley. Lake Volta has a second and smaller clupeoid, *Cynothrissa*, related to the *Limnothrissa* of Lake Tanganyika and now also of Lake Kariba; but *Cynothrissa* feeds on smaller fish, and especially on the young of *Pellonula*.[4]

The swing over in the population of the lake from predatory and insectivorous fish to herbivores is another factor in the great productivity of Lake Volta. Commercial fishing is chiefly by seine and gill net. The two clupeoids are caught at night with the aid of bright lights.[5] So far, Lake Volta has been spared infestation by floating vegetation.

Lake Kainji on the Niger

The River Niger has also been dammed for hydroelectric generation at Bussa. The new Lake Kainji, formed by the dam, is very different from both Lake Volta and Lake Kariba. The resident time is only 76 days, the mean depth is 11 metres, with a 10 metre draw-down, and the lake has an area of 1,250 $km.^2$. Because of the short resident period of the water in the dam, Lake Kainji keeps, so far, much of the character of the river. Nevertheless, during the first two years, a semipermanent thermocline developed during the months of March to May, when the rainy season began. The bottom water became deoxygenated, and H_2S was detected at a depth of fifty to sixty metres. The water of the lake tends to be turbid with suspended particles of clay and mica, and this causes poor light penetration and a low primary plant production of 0.16 to 2.7 $mg.C/m^2/day$.[6]

Plants and Fish Species

Littoral plant communities soon established themselves, and were dominated by the knot grass *Polygonum senegalum*, and the true grass *Echinochloa pyramidalis*, which tended to form dense semifloating beds. These have since been reduced by the long periods of draw-down. The clearance of the lake bed before impoundment was thorough, and only about thirty per cent remained uncleared.

The water cabbage, *Pistia stratiotes*, has always been present in the River Niger, and its pale green cups floating downstream lighten the dark aspect of the Niger Delta. It is disliked because it can harbour *Mansonia* mosquitoes. But it is not a nuisance on the scale of *Eichhornia* or *Salvinia*, and Lake Kainji has not yet suffered from weed pests. The regular daily winds disperse any mats which form, and the short resident period of the water tells against the growth of a weed mat. In fact, both *Pistia* and the grass *Echinochloa* had disappeared from the lake by July 1970; the dam was closed in August 1968.[7]

The lake has a rich algal flora, with the green algae *Volvox* and *Eudorina*, and the blue-green algae *Anabaena* and *Microcystis*. Aufwuchs grows luxuriantly. This abundance of algal food encourages a herbivorous diet among those fish species with a general diet; the well-known fishes *Citharinus* and *Distichodus*, the Cyprinid fish *Labeo*, and even the Characinid fish *Alestes*, have all become predominantly herbivores in Lake Kainji. *Citharinus* in particular adapted very quickly to an algal and vegetarian diet, and grows faster in the lake than it previously did in the river.

The zooplankton is usually dominated by Cladocera, but Rotifers are abundant in July. Copepods and other crustacea are carried into the lake by floods; a new Cladoceran, *Bosminopsis dietersi*, is now becoming dominant.

The fish community is changing in composition, and in abundance.[8] At first, the Citharinidae were the chief fish caught, and in 1969 comprised 37 per cent of the catch, the Characins only 16 per cent. By 1972, this position was reversed, and now the Characins comprised 57 per cent of the fish caught, and the Citharinids only 9 per cent. The most important of the Characins are the species of *Alestes*, which, as in Lake Kariba, came to dominate the surface waters as plankton feeders. Tiger fish were present as predators. There are two important species of clupeoids in Lake Kainji, namely *Sierrathrissa* and a larger species of *Pellonula*. These fish are widely distributed over the open lake and are both small, the largest species only reaching 6 cm. *Tilapia*, one of the most important fishes of the Niger, has not established a large population in this lake, as it has in Lakes Kariba and Volta.

Decline in Fish Populations

But the fish population has been declining since the dam was closed. Total fish production fell from 17,000 tons in 1969 to 11,000 tons in 1972, and the catch per unit fishing effort has fallen from 37 lb. in 1969 to 4.9 in 1972. With this, the earnings of the fishermen have fallen, in spite of a doubling of the price obtained for their fish. The hope for the future is that a large population of pelagic fish will develop in the surface waters of the open lake, and support a fishery not only for themselves, but also for a fauna of predatory fish such as the tiger fish.[9] Secondly, with the growth of an aufwuchs and the development of the lake margins, a large development of the *Tilapias* may give a good and valuable fishery.

Fish catches tend to be best at draw-down and poorest at reservoir high level. The simplest and probable reason is the concentration of the fish population in the smaller volume of water, and its dispersal at high water, which has four times the volume of the lake at low water. During the low water period, catches comprise 18 per cent of tiger fish and 13 per cent of Nile perch; evidently here too the predators have the upper hand at low water. River conditions still control the times of spawning, for the most important species of fish still spawn as the water level rises, as they did in the pre-impoundment river. But other fish, such as the *Tilapias*, spawn the year round.

As a contrast to these tropical dams are some Siberian artificial lakes. The Soviets have been active in hydroelectric development, and claim a capacity of power station dams superior to that of the USA.

Siberian Reservoirs

Two large reservoirs in Siberia may be compared, namely, those on the Rivers Yenisei and Ob respectively. These are in one of the coldest parts of the world, and are regularly frozen over for six or seven months of the year. They are of poor natural productivity; but soon after the closure of the dam, the fauna quickly changed from a riverine to a lacustrine, and the riverine fauna of insects disappeared within a few months.[10] Zooplankton became many times as abundant as it had been in the river, where it was only weakly developed, at a biomass of about 0.046 $gm./m^3$. In the Novosibirisk Reservoir on the River Ob, the resident period of the water is short, and rotifers comprised 70 per cent of the zooplankton biomass, as it had in the river before impoundment. But towards the dam, rotifers diminished to only 2.5 per cent of the biomass, and crustacea comprised the rest of a biomass, which had increased in the impounded lake to 1.03 to 2.87 gm./m^3.

But in the Krasnoyarsk Reservoir on the River Yenisei, the water has a long period of residence, and Cladocera and Copepods soon became abundant and dominant in the zooplankton, while rotifers declined to 4 per cent. The total biomass of the zooplankton increased to 0.75 gm./m^3 in 1967.

Productivity of the Reservoirs

The ice cover, which may be 0.8 to 1.5 metres thick, does not prevent a vigorous algal growth, and in these reservoirs the phytoplankton soon became far richer than it had been in the rivers. In the River Yenisei, for example, the phytoplankton biomass never exceeded 1.3 mg./l., nor 4.0 in the Ob. Yet in the reservoirs formed on these rivers the biomass of the phytoplankton rose to the order of 14 mg./l., with a useful distribution in depth. When intense water blooms of blue-green algae developed, the phytoplankton biomass rose to 440 mg./l.

In both reservoirs, the zoobenthos (animals living on or in or close to the bottom) was poor in biomass at first, and this situation was worsened by the breeding season of the Chironomid larvae, which metamorphosed and took wing at the height of the feeding season for the fish and other predatory inhabitants of the lakes. There was also a development in these reservoirs of organisms, such as large mollusca, which, as in the Thames at Reading (Chapter 3), are of no use to the fish and must form a blind alley in the food chain.

Successful Transplantation of Food Organisms and Fish

Because of this, the Russian biologists have studied the transplantation and acclimatisation to these reservoirs of food organisms of known high nutrient value, following up the great success in transplanting the marine invertebrates *Nereis* and *Syndesmya* into the Caspian Sea as food for the population of sturgeons which live there. It was known that in the reservoirs there was an abundant phytoplankton, and also detritus rich in bacteria, both valuable food reserves for organisms able to use them. Such introductions are best made at an early stage in the maturation of a reservoir, for the fish population will then still be scanty, and the existing species composition of the invertebrates poor.

During the years 1947-67, forty-nine species of invertebrates have been introduced into reservoirs in the European part of the Soviet Union. These include four species of polychaete worms, ten of molluscs, one species of decapod crustacean, nine Mysids, seven cumacean crustaceans, and eighteen amphipods. Mysids and Gammarids from the Caspian, and specialised gammarids from Lake Baikal, have proved to be especially suitable and successful as transplants, and their successful acclimatisation has much improved the food supply for the fish in the reservoirs.

Further, these introductions seem to have had no adverse effect on the original benthic fauna. In a number of reservoirs, those fish which feed on the benthos have much increased in abundance, as a result of these measures. More recently, fastgrowing and efficient plankton feeding fish from the Amur River of Siberia, namely the silver carp, *Hypophthalmichthys molitrix*, and the Big Head, *Aristichthys nobilis*, and the vegetarian fish, *Ctenopharyngodon idella*, also from the Amur, have been induced to breed in the Soviets by hypophysisation and transplanted to many of the reservoirs. They are very valuable food fish.

In the two Siberian reservoirs mentioned above, it was recommended to transplant the crustaceans *Mysis relicta* and *Pontoporeia affinis*, which are the mainstay of the fish in the Great Lakes of America (Chapter 5), into the Novorossisk Reservoir; and the Lake Baikal Copepod *Epischura*, and the Copepod *Limnocalanus macrurus* from the Yenisei River, into the Krasnoyarsk Reservoir.

Fish Population of the Reservoirs

The fish populations of the two Siberian reservoirs consist of the original river fauna and some introductions. In the Novorossisk, the original fish included a population of sturgeon and of *Stenodus nelma*, fish which still breed in the riverine part of the reservoir. Introduced fish include the bream (*Abramis brama*) and the perch pike (*Lucioperca*). Other attempted introductions such as the *Coregonus peled*, a successful plankton feeder of the West Russian and Finnish lakes (it occurs as an Arctic relic in our Windermere), failed, as did the introduction of the common carp, though this is a native of many lakes in Siberia.

In the Krasnoyarsk Reservoir, the best of the original fish was the sturgeon, and the several successfully introduced fish included the omul (*Coregonus autumnalis*) of Lake Baikal, the *C. sardinella*, and the bream. The original predatory fish of the river, the pike and the pike perch, at first increased in abundance in the reservoirs, but now have been superseded by the successful perch and bream.

Commerical fishing is being organised in these reservoirs, and in the Novorossisk Reservoir the rate of fish production is about 5 to 8 kg./ha./annum. This may be compared with 40 kg./ha./annum in Lake Volta; but the Siberian reservoirs are in a region of intense and prolonged cold, and these conditions set a limit to the number of generations bred annually by the most important planktonic and benthic invertebrates, and to the rate of growth of these and of the fish which feed on them.

However, it will be seen that, in spite of such contrasting conditions, these Siberian reservoirs follow the same changes in flora and fauna, following impoundment, as reservoirs in a tropical climate; there is a great increase in biomass, a decline in the predators relative to the non-predators, and a speedy replacement of the riverine with a lacustrine type of fauna and flora.

Fertile Reservoirs

Some reservoirs in Soviet Turkmenia and in the southeastern United States are interesting because, although their waters are oligotrophic, or very poor in plant nutrients, they are highly productive and support very dense stands of large water plants.[11] In one Turkmenian reservoir, the Khalif, the dry weight per hectare of aquatic macrophytes was as follows:– *Ceratophyllum demersum*, the pondweed, 24 metric tons, *Myriophyllum spicatum*, the water milfoil, 5.5, *Typha angustata*, the reedmace or catstail, 26, and *Phragmites communis*, the reed, 41 metric tons per hectare. Phytoplankton is scarce, as it also is in oligotrophic lakes in the southeastern United States; in both cases, these rooted water plants are able to use nutrients from the usually anaerobic bottom muds for their very abundant growth. It has been suggested that one means of reducing plant nutrients and so diminishing eutrophication in water bodies would be to grow harvestable aquatic plants, preferably plants useful to man.

Management of American Reservoirs

Reservoirs in the United States are managed for sport fishing, and about 25 per cent of all freshwater fishing in the country occurred on man-made lakes. On big impoundments, 100 million man-days are spent yearly, and the mean sport harvest of fish from 121 reservoirs investigated was 22.6 lb./acre in 1968 (20.1 kg./ha.). Reservoirs are the only type of fish habitat able to undergo extensive expansion in the near future.[12] Commercial fishing is allowed only for the purpose of controlling the numbers of so-called ‘rough fish.’ The mean commercial harvest of these commercial fish was about 10.2 kg./ha. in forty-six chiefly southern reservoirs.

The various management measures include stocking with forage or predatory fish to keep a good balance between the two: and to introduce new and desirable species. In waters where trout are heavily fished but do not reproduce well, massive stocking with hatchery reared fish may be done. Brown, brook, lake and rainbow trout are stocked, with rainbows giving the highest returns.

Of the two shads which are valuable forage fish, the threadfin (*Dorosoma petenense*) is superior to the gizzard shad (*D. cepedianum*), partly because of its feeding habits, but also because it grows more slowly, thus being available to predators for a longer period than the faster growing gizzard. So measures are taken to reduce the population of the latter, both by rotenone poisoning and by using the period of draw-down of the reservoir to remove them by licensed commercial fishermen. During the draw-down, also, a quick growing crop such as rye, sudan grass and other plants may be sown in the exposed mud. The result is clearer water and increased fishing success. The use of the draw-down is a potent management tool for controlling the fish populations of reservoirs. The suppression of the gizzard shad results in an increase in the abundance of the threadfin shad, with immediate benefits to the desirable game fishes such as white bass, crappies, catfish, striped bass and trout.

There are many other such management measures, including timber clearing. Some timber left standing in the water of course dies, but, as in the African reservoirs already mentioned, grows aufwuchs, harbours useful insects as fish food, and deters speed boats and waterskiing. One record suggests that in flooded woods, the angler harvest was 3,054 lb./acre (2,718 kg./ha.), while in the open water it was 113 lb./acre (100 kg./ha.). Though the emphasis is on recreational fishing, the totals caught add up to an important fishery. An estimate gives a total of 90,000 metric tons of fish caught from the 3.7 million hectares of American reservoirs.[13]

Fish Migrations and Fish Passes

It has long been known that obstructions across rivers, for whatever purpose, would hinder or prevent fish from reaching their spawning places higher up the rivers. The famous Hell's Gate in the Fraser River of British Columbia, was a naturally produced obstruction which had serious effects on the important salmon fishery of that river system until an efficient fish pass enabled the salmon to ascend beyond the obstruction. There are many types of fish pass or fish ladder; but most consist of a series of small ponds down which a stream of water flows. The effect is to break the barrier of the main dam into a large number of small dams each of a gradient which fish can negotiate. Other types are fish lifts over the obstruction, and can be automatic. There is pressure to construct fish passes wherever obstructions are put across rivers. But as these are expensive, care must be taken to see that there is a real need for such as pass, and that there are fish, especially valuable fish, which must pass upstream of the barrier if their breeding is to be successful. The sight of fish milling about in the turbulent water below a dam does not necessarily mean that these fish are striving to pass up. Below a weir or dam, the water is in rapid motion and is well-oxygenated; it is an area of locally increased aeration and fertility. But it has been found that these fish may show no sign of maturation; they are feeding, not breeding. The big Gebel Aulia Dam across the White Nile above Khartoum had a fish pass incorporated in the design, but few fish use it. In practice, fish passes are only justified for valuable fish such as salmon and, in Russia and Rumania, sturgeon. A portable sectional aluminium fish pass has recently been invented by an American fisheries engineer, which can be air-lifted to a site and assembled there. This valuable device has made many miles of spawning beds accessible to salmon in an Alaskan watershed.[14]

Where fish passes are not practicable, artificial breeding may be resorted to. The many hydroelectric dams across the Swedish rivers, and the important timber industry there, have wrought havoc with the salmon spawning grounds. So the adult salmon ascending from the sea are caught alive and held until spawning time, when they are stripped of their eggs and milt in hatcheries. The young salmon are reared to the smolt stage, when they are released to return to the Baltic Sea. Many years of experience have made these hatcheries expert, so much so that the proportion of artificially reared smolts which later return to the river as adults is as high as was found in the previous natural spawnings. The River Volga in Russia is so completely canalised that it is most economical to catch the adult sturgeon on their spawning migration, and strip and fertilise the eggs artificially, and thus keep this valuable fishery flourishing.

The seaward descent of the young salmon and sturgeon is easier, for it has been shown that they can pass through the turbines with a reasonable rate of survival, and over spillways with little harm.

Eel Migrations

It has been said that the salmon and (in Russia) the sturgeon are the only fish for which the expense of fish passes are justified. Young migrating eels of the species *Anguilla nebulosa* have been found in the pool at the foot of the 400-foot dam wall at Kariba. But the surprising thing is that, somehow, these young eels manage to surmount this vast obstacle, and are forming a population of eels in Lake Kariba in depths of twenty to forty-five metres.[15] This eel migrates to a height of 2,400 metres in the Athi River of Kenya,[16] and in the course of this long pilgrimage they surmount many rapids and waterfalls, including at least one with a height of twenty-five metres, climbing up the water-splashed sides of the falls. The same species has been followed in its migration up the large Indian River Godavari, and has been photographed wriggling up the damp sloping concrete of the weirs at night, seeking places where the current was sluggish or where the concrete was kept wet with spray. At a dam eighty km. from the sea, the young eels had a length range of 48 to 58 mm.; at a dam much further upstream, 725 km. from the sea, these young eels were found still ascending, but here they had a length of about 152 mm. As this is about one year's growth, the eels are evidently making their way slowly upstream, growing as they go.[17] This is in line with the upstream migration of the European eel, *Anguilla anguilla*, which negotiates obstacles in the same way, sometimes forming a rope of living and dead elvers up which the later arrivals climb. In this eel, also, the upstream migration may continue for several years, the biggest fish going furthest upstream, and these bigger fish are all females. It has recently been shown that the upstream migration is so slow, in the Irish River Shannon, that the upstream lakes of that river system are much underpopulated with eels.

In Kenya, Kikuyu fishermen used to come to the mountain streams with traps, to catch the big eels on their downstream migration as mature silver eels. Smoked eel, of course, is a luxury food; and if a fishery for eels can be worked up in Lake Kariba, there would surely be a good market for smoked eels in the cities of south and east Africa. It would not be difficult to transport live elvers from the foot of the dam to the lake; eels travel well in damp cloth trays, and can be closely packed. Hundreds of millions have been caught in the River Severn, and flown to Germany and Poland to stock their inland waters, which do not naturally get a sufficient supply of elvers.

Dam Building and the Spread of Disease

The building of these massive dams, all major engineering works, can result in the spread of disease. During construction, many thousands of people are attracted to the site; these include not only the skilled engineers and technicians, who will know the rules of hygiene, but also great numbers of local folk and their dependents, many of whom are certain to be infected with diseases which are transmitted by water. Though a township of good standard may be put up for their accommodation, there often arises alongside a shanty town where unemployed labourers, would-be labourers, hucksters and vagrants doss down. They will have gathered from a wide area, and their diseases will be quickly spread in such crowded conditions.[18]

Conditions may favour the spread of malaria, for malarial mosquitoes will find plentiful blood meals among the labour force, many of whom will harbour the malarial parasite in their blood. Before these diseases were understood, there were terrible mortalities among the crowded labour forces at the Panama and Suez Canals: both were nearly abandoned for this reason. The damp soil will also favour the spread of the debilitating hookworm parasite.

Another serious disease favoured by dam building is onchocerciasis, caused by the nematode worm *Onchocera*, which is transmitted from one human to another by the bite of blackflies, *Simulium* spp. These have been mentioned in Chapter 2, as inhabitants of running streams; they have a painful sting, and are a considerable nuisance in their own right. They can find in dam sites many favourable places for breeding, such as spillways and diversion channels. Blackflies were a serious obstacle to the building of the Owen Falls Dam in Uganda; but, luckily, DDT had then become available to abate the nuisance. This benefit of abolishing the blackflies went further; for wide areas of land on either side of the Nile below the Owen Falls were uninhabited because of the blackfly scourge; with their elimination this neglected land was settled by farmers and brought into production.

The *Onchocerca* parasite is liable to enter the eye in the course of its wanderings about the body of an infected person, and frequently results in partial or total blindness.

A third disease associated with water, and so with the crowded populations of the dam builders, is Bilharzia, or schistosomiasis. It is now recognised as a dangerous disease. An infected person passes the eggs of the parasite out in his urine or faeces, and the eggs hatch out in water into a mobile *miracidium* larva. This swims about until it meets a watersnail of certain genera which act as intermediate hosts; then it penetrates into the tissues of the snail and lives there, finally turning into a mass of larvae called *cercariae*, which are released into water to swim until they can find the bare foot or leg of a human. They can burrow into the skin, get into the bloodstream, and thence into the big liver vein, where they mature, sending out their eggs in myriads. The eggs find their way out through the bowel or the bladder into water to repeat the cycle. The patient becomes anaemic and feeble, unfit for work, and in such an anaemic state is vulnerable to other diseases. Because of the millennia of irrigated cultivation, with its prevalence of water and damp soil, and the primitive sanitary habits of the peasants, Bilharzia has always been endemic in Egypt, and unfortunately there has been a great expansion of the disease since the Aswan High Dam made greatly increased areas of irrigation possible. Unless it can be curbed, many of the benefits of the dam will be lost.

A recent article by an Egyptian author points out, rather testily, that the eggs of *Schistosoma* have been found in mummified bodies thousands of years old, so that this disease has been prevalent during most of Egypt's history. But this handicap did not prevent the construction by the peasants of the stupendous monuments and public works which survive to this day. However, it is sad news that Bilharzia has now appeared in Lake Kariba.[19] Some sixty-eight per cent of the school-children at Siavoga, a new lakeside township, are infected with *Schistosoma haematobium*, and sixteen per cent with *S. mansoni*. The snail vectors live and breed in the mats of *Salvinia*, which may drift about the lake, and carry the infected snails with them. These breeding sites will have to be treated with molluscicides, and cleared out wherever possible. It may even be desirable to introduce a snail-eating fish, of which at least one has a high food value.

NOTES

1. Petr, T. 1972. 'Benthic fauna of a tropical man-made lake.' *Arch. für Hydrobiol.* 20 (4): 484-533.
2. Viner, A.B. 1970. 'Hydrobiology of Lake Volta.' *Hydrobiol.* 35: 230-48.
3. Petr, T. 1971. 'Lake Volta, a progress report.' *New Scientist* 49: 178-82.
4. Petr, T. 1968. 'The establishment of lacustrine fish populations in the Volta Lake in Ghana, 1964.' *Bull. IFAN* XXX Ser A (1).
5. Reynolds, J.D. 1970. 'Biology of the small pelagic fishes in the new Volta Lake.' *Hydrobiologia.* 35: 568-603.
6. Anon. 1971. 'First Symposium on Lake Kainji.' *Afr. J. Tropical Hydrobiol. and Fisheries* 1 (1): 67-8.
7. Imevbore, A.M.A. 1971. 'Floating vegetation of Lake Kainji.' *Nature* 230: 599-600.
8. Turner, J.L. 1970. 'The fish population of the newly impounded Kainji Lake in Nigeria.' *J. West Afr. Sci. Assn.* 16 (1): 49-58.
9. Lelek, A. 1972. 'Kainji Lake Research Project.' UNDP of F.A.O. Rome, 1972.
10. Popova, T.G., Vershinin, N.V., Blagovidova, L.A., Setsko, R.I., and Chaikovskaia, T.S. 1972. 'Biological characterstics of the first reservoirs on the Ob and Yenisei Rivers.' *Verh. int. Verein. theor. angew. Limnol.* 18: 872-6.
11. Kogan, Sh. I. 1972. 'Some biological peculiarities in reservoirs of south Turkmenia.' *Verh. internat. Verein. Limnol.* 18: 867-71.
12. Jenkins, Robt. M. 1972. 'Reservoir Fish Management.' *A Century of Fisheries in North America.* Sp. Pubn. No. 7. Amer. Fish. Soc.
13. Jenkins, R.M. and Morais, D.I. 1971. 'Reservoir sport fishing effort and harvest etc.' in *Reservoir Fisheries and Limnology,* Ed. Gordon E. Hall. Sp. Pubn. No. 8. Amer. Fish. Soc.
14. Eicher, G.J. 1972. 'Fish Passages.' in *A Century of Fisheries in North America.* Sp. Pubn. No. 7. Amer. Fish. Soc.
15. Balon, E.K. 1971. 'Notes and Comments.' *Afr. J. Trop. hydrobiol. Fish.* 1 (1): 76. 1971.
16. Frost, W.E. 1955. 'Observations on the biology of Eels [*Anguilla* spp] of Kenya Colony. E. Africa.' Colonial Office Fish. Pubns. 6. 1955.
17. Ibrahim, K.H. 1961. 'Upstream migration of elvers of *Anguilla nebulosa* (= *bengalensis*) over first anicut of the River Godavari.' *J. Bombay Nat. Hist. Soc.* 58 (3): 810-2.

18. Waddy, B.B. 1966. 'Medical problems arising from the making of dams in the tropics.' *Man-made Lakes.* London, Academic Press.
19. Hira, P.R. 1969. 'Transmission of Schistomiasis in Lake Kariba.' *Nature.* 224: 670-2.

Chapter 8 THE POLLUTION OF WATERS

> I was commanded, with the Water-Bailey,
> To see the rivers cleaned, both night and daily,
> Dead hogges, Dogges, cates, and well-flayed carrion horses,
> Their noysom Corpses soyld the water-courses,
> Both Swines and stable dung, Beasts guts and garbage,
> Street durt, with Gardners Weeds and Rotten Herbage,
> And from those Waters' filthy putrefaction
> Our meat and Drinke were made, which bred Infection,
> Myself and partner, with cost paines and Travell,
> Saw all made clean, from Carryon, Mud, and Gravell,
> And now and then was punisht a Delinquent,
> By which good meanes away the filth and stink went.
>
> John Taylor, 1644.

Thus far, water has been followed along its way from the streams into the rivers, from the rivers into natural and artificial lakes, and thence to its first contact with the sea in the estuaries. This would seem to be the appropriate place in which to write of pollution, which is common to all aquatic environments, freshwater and marine.

The Shorter Oxford Dictionary defines 'pollute' as: to make physically impure, foul, or filthy; to dirty, stain, or taint. Water is particularly liable to pollution, often naturally, but, in the present industrialised urban society, by deliberate and usually preventible fouling. Because water flows, offal dumped in it is carried away to afflict people downstream; lakes and even large ponds can accept a lot of rubbish and offal without much of it showing. As to the sea, its volume is so great that very great amounts of all kinds of wastes can be absorbed. This transport and disposal of offal is free: and towns and villages have always been built beside water not only because of an assured supply of water, but also because of this free waste disposal facility.

The doggerel verse quoted shows that some efforts were made in the past to control pollution; but in modern times pollution has become a major issue, with powerful lobbies in the legislatures and ever stricter restraints.

Pollution can be conveniently grouped under seven heads, namely: silt; organic matter; chemicals; herbicides and insecticides; oils; radionuclides; heat and general rubbish.

Pollution by Silt

Given sufficient time, the tooth of running water is sharp,[1] and it is always gnawing away the land through which it flows. The finer particles are moved in bulk, for long distances. They may settle and smother all plant and animal life in bad cases; but though some fishes, and no doubt many invertebrates, can tolerate high turbidities for short periods, more than 100-200 parts of suspended material per million of water can be harmful or fatal.[2] Landslides, collapses of banks in spates, and the grinding of ice expose bare soil and can lead to damaging siltation. But not all this siltation is wholly harmful. For at least 7,000 years, the Nile flood has annually brought not only the water to irrigate the harvests of the husbandmen but the fertilising silt to nourish them (p. 32).

Much siltation is man-made. Mineworkings, tunnelling, quarrying, road making and such works are all liable to lead to the choking of the water courses. The felling of forests, the clearing of bush and burning or overgrazing of grass cover, can all lead to the carrying off of loose soil no longer bound by plant roots and detritus. The Ord River of north-west Australia has now been dammed as part of a major irrigation scheme. When in flood, it carried a large silt load which, unchecked, could settle in the reservoirs and shorten their useful lives. An increase of siltation was traced back to overstocking, and so to overgrazing of the poor pastures of the Kimberleys, which exposed the soil to the torrential rains of that region. A major catchment rehabilitation scheme was undertaken by the West Australian Government, and the position is much improved.

China clay workings are responsible for large loads of fine white silt which are carried out to sea and settle down in the bays. I have seen Plymouth Sound a brilliant green colour with suspended china clay. This white silt chokes off most life on the sea bottom, producing near-sterility. But further offshore, where the rate of deposition is less, a rich fauna of the *echinocardium/filiformis* type developes, in contrast to a zone closer inshore, where near sterility prevails.[3] This china clay fauna of polychaetes and lamellibranchs may actually improve the bays as feeding grounds for bottom feeding fish, an unexpected result. This is not to be taken as a reason for dumping these wastes in the sea; it should rather be taken as an instance of how nature can cope with this kind of abuse.

Pollution by Organic Matter

This is perhaps the most common type of pollution, and its effect is largely indirect. Organic matter in water is oxidised by bacteria; as the dissolved oxygen is thus used up, several consequences follow. First, fish and many other organisms may die of suffocation, and those fish which need most oxygen, such as the active members of the salmon family, will die first. Many organisms can tolerate for a while extremely low oxygen concentrations, but these will also die in the total absence of oxygen. The only aerobic life to go on in such water will be those organisms able to breathe atmospheric air, fish with lungs or accessory air breathing organs, or insects with similar adaptations.

Secondly, when all aerobic life is killed, a new flora of anaerobic fungi and bacteria takes over and flourishes. These plants live without oxygen, by reducing the organic material which does contain oxygen to fermentation products, many of which have a foul appearance and smell. The organisms include the dirty grey sewage fungus, and the products include the foul-smelling and poisonous gas sulphuretted hydrogen. Even sawdust in the streams beside sawmills will blacken and stink.

Luckily, recovery can be rapid when dilution takes place with well-oxygenated water, in which green plants can produce plentiful oxygen in sunlight (p. 3), and which flows over cascades or turbulent eddies which mix in oxygen from the air.

Natural Pollution

Nature is responsible for this type of pollution on a massive scale. In swamps, for example, an excess of decaying organic matter derived from dead vegetation can remove all life-supporting oxygen from great volumes of water, and when this water drifts about or is carried into rivers, massive mortalities of fish follow. Great volumes of marsh gas, or methane, may also be formed; and Debenham describes how, in the Bangweulu Swamps of Central Africa, his motor launch was followed by flames three feet high, as the methane released by his passage was ignited by the engine.[4]

Another common case occurs in rivers at the time of low water in the summer: decaying organic matter on the banks and in the stream may be swept downriver in the spate following a heavy rainstorm, and by robbing the water of its dissolved oxygen may kill fish and other organisms. For twenty miles downstream, thousands of salmon, trout, roach, chub and dace may be killed.[5]

Natural pollution can also occur in lakes and ponds when an excessive growth of the phytoplankton, and especially of the blue-green algae, produces a complete deoxygenation of the water at night, when photosynthesis is at a standstill (p. 4). Many fish may be killed by suffocation, and such kills are called 'summer kills,' when high temperatures, abundant plant life and organic matter, overcast skies, and absence of a stirring wind, combine to remove all dissolved oxygen. Relief comes when bright sunshine and stirring breezes reintroduce dissolved oxygen. This is one of the occupational hazards of tropical and subtropical fish farming. To keep up some supply of oxygen and so save a crop of valuable fish, it may be necessary to create fountains with portable pumps, or, in small ponds, thrash the water with long bamboos, a desperate measure indeed, but which often succeeds.[6]

So-called 'winter kills' are equally damaging. When ice has sealed off the water from the air, green plants can still produce enough oxygen to support animal life if the sunlight can penetrate the ice. But if a blanket of snow prevents this, decaying organic matter may expose animals living beneath the ice to suffocation.

Man-Made Pollution

But man-made pollution attracts attention nowadays and is one of the chief environmental problems. The growing human population is greatly accelerating the accumulation of its waste products, so much so that the doomwatchers foresee mankind dying out in its own noxious wastes like any culture of yeast or bacteria. But the modern degree of water pollution is only relatively worse than in earlier years. The doggerel rhyme quoted above shows how sewage and offal were disposed of at Oxford; but at least there were regulations and an attempt to enforce them. In the Middle Ages, London dumped its sewage into the Thames or the Fleet Ditch, or into the gutters. No wonder life then was short and disease rampant. Things slowly improved, but in the 1850s the London sewage drifted upstream on the tide and made conditions so foul that Parliament had to adjourn and Queen Victoria and the Prince Consort had to abandon a projected river cruise. The modern Thames has known nothing so bad as this.

The commonest organic pollutant is sewage, but other pollutants are the washings and wastes of milk, cheese, and starch and sugar factories, distilleries and meat plants. Feed lots, farmyards and silos in the country also contribute organic pollutants. In winter, when rivers are usually in full flow and cold enough to dissolve a large amount of oxygen, sewage can be accepted, but in summer, when rivers are low and at high temperature the discharge of oxygen robbing wastes can be fatal to aquatic life over long distances.

One method of sewage disposal is to pump it through long pipes far out to sea, in areas where the currents will disperse it for dilution and speedy destruction. Estuaries have long been 'the septic tanks of Megalopolis.'

But there are now well-established methods of treating such putrescible material. In modern sewage works, the solids in the sewage are allowed to settle out to undergo anaerobic fermentation. A large part of the carbon is set free, bound to hydrogen in the inflammable gas methane. This gas may be collected, and used or sold as fuel. The spent sludge has a high fertiliser value. After the settling out of the solids, the clarified liquor may be sprayed down a trickle filter, where the remaining organic matter is oxidised by a bed of clinker well infected with bacteria, protozoa, fungi and algae, together with fly larvae and worms. The film of bacteria and fungi break down the organic matter into simpler harmless forms, and the insects and other animals clear away excessive growths of the bacterial film. The liquor now contains very little putrescible matter and can safely be discharged to natural waters.

The fuel crisis of 1973-4 has led to an inventory of possible alternative sources of energy, and included among these was the anaerobic fermentation of town refuse as well as sewage for the recovery of methane. More recently, Californian biologists have suggested that brown seaweeds should be cultivated as raw material for methane fermentation. The coastal seaweeds are extremely productive, especially when constant cropping prevents autoshading and diminished growth (p. 127).

London can take a civic pride in its recent record. Though, by 1850, all commercial fishing in the London area had ceased, at about that time the authorities installed a system to collect all sewage effluents and discharge them far down the estuary. In the 1890s fish began to return to the river; but the rapid growth of the population led to overloading. By 1953, anaerobic conditions prevailed for much of the year for at least twenty miles below London during summer.

In 1960, the London County Council rebuilt and modernised the treatment plant, and since then over fifty species of mainly marine fish have returned to the lower estuary, with freshwater species upstream of London. In 1972, the first sea trout was caught, and a prize was offered for the first salmon to be caught in the Thames for well over a century. That prize may be claimed within a decade. Amid the outcries of the doomwatchers, here is a small but clear call of success. But the estuary of the Rhine is still grossly polluted, so much so that even the very pollution-tolerant *Tubifex* worms are endangered.

A notable natural purification of water takes place in lakes and reservoirs. For example, in an English storage reservoir, the inflow from the slow-flowing River Stour (p. 27) had a titre of the intestinal bacterium *E. coli* of 5,470, but at the outflow, of 28/100 ml. In a storage reservoir in France, water returned to the River Seine to top up the low summer flow is completely free of pathogens. This is the general experience.

Eutrophication

Sewage and other putrescible matter contains, besides the carbon compounds, many other substances which contain phosphorus and nitrogen, and trace quantities of metals. These are all valuable plant nutrients; so they are released when the material is oxidised, and support a vigorous growth of plants. The suspended sewage solids in the River Kennett (p. 26), when mixed with the Thames water, oxidise at the expense of some of the oxygen the latter contains and release abundant nutrients which support the rich phytoplankton of the main river.[7] Such an addition of nutrients is called eutrophication, and it is on the increase, as both sewage derived and agricultural fertilisers accumulate in natural waters. It might be said that just as a weed is a plant in the wrong place, organic pollution is a fertiliser in the wrong place.

But there are many advantages as well as disadvantages in eutrophication. Clarified sewage effluent is a good pond fertiliser, and many cities, of which Munich is the best-known, purify their sewage by passing it through large, shallow ponds. In oxidising in such favourable conditions the sewage releases so much plant nutrients and detritus that a food chain is set up which leads from algae to the water flea *Daphnia* which is the preferred food for the commercial carp. Fish farmers rent these ponds with profit, for the fish crops equal those obtained by expensive fertilising and foddering in the usual production ponds. Raw sewage is much used in eastern Asia as a pond fertiliser, giving large crops of fish.

Modern 'soft' detergents are a new, additional source of the key nutrient phosphorus in sewage. The treated sewage of the London conurbation supports, in the southern North Sea, a year-round crop of phytoplankton, and nourishes an enhanced fish population as compared with adjacent unfertilised areas of sea. It has been calculated that, in 1948, about 300,000 tons of commercial fish were harvested from the southern North Sea, and of this about 100,000 tons could be attributed to the fertilising effect of the London sewage. It has recently been estimated that sewage and run-off contribute 1,548 tons of nitrogen and 256 tons of phosphorus daily to the North Sea.[8]

But, generally, eutrophication sets problems, for it can change the flora and fauna of lakes, for example, and by stimulating the massive growth of aquatic vegetation may lead once more to oxygen depletion at unfavourable times. Lake Erie, which is badly affected, has been much changed by eutrophication, from a lake supporting the so-called clean water species such as Cisco, whitefish and blue pike, to one supporting more pollution-tolerant species such as white bass, sheepshead and carp.

Pollution by Chemicals

In this class of pollution, the damage is direct, since poisons are involved. The substances vary as much as industry itself varies; phenols, cyanides, arsenic (especially from sheep and cattle dips), chlorine, slaked lime, and compounds of many heavy metals. Many modern chemicals are very poisonous; the case of a chemical discharged into one of the tributaries of the Rhine is well-known; it killed thousands of fish all the way down the main river. On a small scale, paint remover tipped into a salmon and trout stream killed all fish there. Cases like these are all too frequent. The only way of treating such poisonous residues is dilution to the point where they become harmless, in the sea or large lakes, for example.

But even there, harm can be done. The discharge of effluents containing heavy metals such as mercury and cadmium into the Inland Sea of Japan resulted in many deaths. Oysters can accumulate heavy metals in their bodies, and as much as 300 parts per million, wet weight, of copper and zinc have been recorded. Cockles, and also some fish, can accumulate enough metals, such as mercury, to make them unacceptable as food. Tunas, which are oceanic fish, are among those in which high concentrations of mercury have been found. It seems likely that declines in the stocks of shrimp and prawns are associated with the discharge of sewage.[8] Seabirds have a tendency to accumulate these metals, because they are at the head of marine food chains.

It has been said that pollution may effect the fortunes of the sea fisheries through its effect on the survival of very young fish. Most of our important fish species spawn close inshore, and their young settle on or near the sea bottom to grow. They may then be exposed to considerable pollution, including heavy metals. Already it is well known that the heaviest mortality in the life of these fish occurs in these and younger stages; anything which enhances this mortality is bound to affect the fish harvests in later years.

Pending more knowledge of the harmful effects of heavy metal pollution, the best policy seems to be to monitor the discharge of potentially toxic metals while an acceptable permissible maximum is being agreed on. Then, the cost of alternative means if disposal has to be considered. It would be passed on to the consumer, who has to decide how much he will pay for a stand against increasing pollution of all kinds.

Insecticies and Herbicides

Persistent insecticides such as DDT now have a worldwide distribution, for they are usually sprayed on crops, and some may be carried upwards and widely dispersed. It is well-known that measurable quantities of DDT have been found in penguins and seals in the Antarctic, remote from any likely places of usage. By means of the food chain, DDT accumulates in the predators. High concentrations have been found in sea birds, for instance, and their fertility is effected, not least by a tendency to lay thin shelled eggs which are easily broken.

DDT and other chlorinated hydrocarbons such as endrin, aldrin and dieldrin are extremely toxic to fish, so that heavy fish kills can follow crop spraying, if some of the spray is blown into the water. Laboratory tests have shown that concentrations of only 0.6 parts per million of endrin, aldrin and dieldrin killed fish respectively in one and a half, five and a half, and more than seven hours.[9] Fishes that survive can concentrate insecticide in their bodies and build up a ten-fold resistance, as compared with unexposed fishes. A case is quoted where a single resistant *Gambusia* fish released enough endrin into ten litres of water to kill five previously unexposed fish in thirty-eight hours.

The organic phosphorus insecticides may be more immediately toxic to fish than the chlorinated hydrocarbons, but are much less persistent and seem to be less dangerous in the food chains.

It seems that we shall have to live with this problem, at least until less toxic insecticides, or alternative methods of control, are invented. The modern high crop yields on which the feeding of a growing human biomass depends are in large part due to the control of plant diseases and pests, and especially of storage pests, by these powerful chemicals.

Herbicides such as diquat and paraquat are not at present regarded as pollutants when used at the recommended dosages.

Pollution by Mineral Oils

This is probably the best-known kind of pollution, thanks in part to the massive publicity given to the 'Torrey Canyon' spill. More or less effective controls limit the pollution of inland waters, though industrialised rivers and canals are sometimes badly oiled. Spent oil from car sumps is often poured down the drains, whence it will appear in the rivers. The effect of such oil spills is less disastrous than might be expected. In a pond contaminated with spent lubricating oils, those insects which had to come to the surface to breathe air were killed (in fact this is one means of controlling the breeding of dangerous mosquitoes), but aquatic clams and snails actually fed upon the oil film and the organisms living on it, passing out the oil in the faeces as little droplets. The harmful effect of the oil diminished in a few days, and the pond was then recolonised by insects.[10]

Oils and oil processing effluents can be held for up to forty days in a system of ponds, and then lose nearly all their harmful components. Such effluents also soon lose their toxicity to fish as they pass along open ditches. But outboard engines can pollute water from underwater exhausts and cause tainting of fish when the fuel use level reaches eight gallons per million.

Oil spills in the sea are slowly dispersed by natural means. The light fractions evaporate, and bacteria slowly oxidise the heavier fractions.

But the heaviest fractions are very slow to disperse; they have been seen far out in mid-ocean, and are only too familiar cast on the beaches, deserving the curses of seaside hoteliers as they are brought indoors on the shoes and towels of guests. The damage done in the 'Torrey Canyon' incident was due to a near panic, as the beaches were fouled just at the beginning of the holiday season. Excessive quantities of detergents were used, to try to clear the beaches, and these did far more damage to marine life than the oil would have done. Even now, years later, the ecological damage then done has not been wholly restored. Browsers like the limpets and Chitons will clean oil off rocks and coral, scraping and swallowing the film of oil and the algae growing on it, and passing the oil out as easily dispersed droplets. Long ago, Orton saw marine animals grazing on the underside of oil films in sea water.

Oil films floating on the sea are harmful to diving seabirds, whose miserable fate is well-known. More recently, seal pups have been found with contaminated fur; and oil films stranded on shellfish beds may smother and certainly taint the shellfish.

International legislation now controls the deliberate discharge of oil at sea, and there have been many successful prosecutions. But, as oil tankers become ever bulkier, with half million tonners on the horizon, the danger of accidental oil spills on a huge scale must be provided for.

Pollution by Radionuclides

The study of radioactivity in water is a very young science, barely twenty-five years old, and the nuclear industry which has emerged since the war is unique, in that protection standards against radioactive effluents were developed before the technology became widespread.[11]

Nuclear power stations need a great amount of cooling water and are situated on rivers or on the coast. The risk of contamination of the environment with radionuclides, well-known to be dangerous since the early researchers lost fingers, and luminous paint workers developed cancers, has resulted in public concern that this new genie be kept under close observation. It is agreeable to read that the routine disposal of radioactive wastes is now well controlled and all the most important public health problems have been met.[12]

In America, the Clinch River, Tennessee, has been studied for twenty years, with the result that the river can be used to dispose of small quantities of radioactive effluent without endangering people dwelling downstream. Britain's nuclear power stations are situated on the coasts, partly because her rivers are small, while there is unlimited cooling water in the sea. The twenty nuclear power stations which are, or will be, established will regularly discharge their effluents into the estuaries and coastal waters, and form one of the largest concentrations of marine radioactive disposal areas in the world: yet careful control, and siting where the natural currents will disperse the effluents, give them only a small significance as pollutants. An exception is Windscale, in Cumberland, situated on the Irish Sea. But here, also, there is the most careful monitoring. At this plant, spent nuclear fuels are processed, and some of the fission products are discharged into the Irish Sea. Some of the Ruthenium 106 becomes concentrated in the alga *Porphyra*, from which the Welsh seafood 'laverbread' is made. But the exposure to radionuclide of the consumers was found to be well within the safety limit.

Similarly, fishermen exposed to the radioactive effluents of the Dounreay nuclear facility, through handling boats and fishing gear, suffer only 2.5 per cent of the acceptable dose rate. At Bradwell, where a nuclear power station discharges effluents which contaminate an oyster fishery, eating the oysters will give the consumer only 0.2 per cent of the permissible dose limit.

The general conclusion seems to be that aquatic organisms will not at any one time contain a great proportion of the radioactivity in the water, and that will be within what are regarded as safe limits.

The present situation with regard to radioactive pollution then seems reassuring; yet public concern is still lively, as may be seen in the treatment of the Japanese nuclear ship 'Mutsu.' For fear of nuclear contamination, Japanese fishermen are barring her re-entry into port. She seems fated, like a nuclear Vanderdecken or Flying Dutchman, to wander harbourless about the ocean.

Thermal Pollution

Hot water is a pollutant because it contains much less dissolved oxygen than water at lower temperatures; and also because, at temperatures above about 37°C, the number of plants and animals able to tolerate the heat rapidly diminishes. Above 42°C, very few can live. This is well known from the study of hot springs; but the problem has become worldwide with the increase in the number and power of electric generating stations cooled by river water.

Much depends on the climate and latitude of the station. In cold climates, the usual rise in temperature of about 8°C to 15°C in river water, after passing through the condensers, does not raise the temperature to near danger point for river life. In Britain and northwestern America, the heated effluent very seldom exceeds 30°C, and then only for a short period. But in the hotter parts of the United States, temperatures may rise to 35°-42°C, though again, the highest of these is not maintained for long. Another important fact is that peak demand for electric power in Britain usually occurs in winter, when water temperatures are low and river flows ample: whereas in America, due to air-conditioning machinery, peak demand is in the height of summer when river temperatures are high, and river flows are at their lowest. American rivers are then drawn upon for cooling at the time when there is least water to do it, and that at the highest temperatures of the year.[13] Though some deoxygenation may occur below some direct-cooled stations in America, the dissolved oxygen very rarely falls below saturation value at those temperatures. America has abundant water, so direct cooling is the rule there. But in the Green River station in Kentucky, one of the largest steam plants in the world, the invertebrate fauna was found to be seriously affected along five miles of the river shoreline, where the temperature exceeded 35°C for long periods. Where outfall temperatures exceeded 40°C, no living invertebrates were found. It was necessary to instal cooling towers of great size to keep the temperature down to 32°C.

By contrast, in Britain there is less river water in relation to need, so indirect cooling by towers, which needs much less water, is the rule. It is found that the passage through these towers improves the chemical and biological quality of the water. Polluted river water is there given rapid aeration; the polluted River Trent in England receives up to 3.9 metric tons of oxygen per day in its passage through the cooling towers of the 2,000MW Drakelow Power Station.

The effect of the hot water has been studied at the Konakov power station, which discharges its cooling water into the Ivanikovo reservoir near Moscow.[14] The heated water is about 10°C above ambient in winter and 4°C in summer.

The zooplankton of the heated area undergoes important changes. Firstly, about sixty per cent of the Daphnias, Cyclopoids and Rotifers, and all the larger zooplankton forms, die on passage through the condensers. The heated water of the reservoir is avoided by Daphnias, Cyclops and Diaptomus species, which are found in the deeper cold water. When there is complete mixing and the temperature rises to 28°-30°C, all zooplankton declined.

The benthos was adversely affected by temperatures of more than 26°C, and its biomass decreased. Below this temperature, at 20° to 25°C, the invertebrate bottom fauna, especially oligochaetes and chironomid larvae, was stimulated and its biomass increased.

The principal fish species were bream, roach and perch, which were little affected by the heated water; but the warmth-loving bleak and rudd flourished. Incidentally, the heated water from power stations is being increasingly used in Russia for the culture of common carp and of valuable Asiatic carps. The warm temperature increases their growth rate and therefore their profitability. In Britain, these experiments are being done in brackish water, with waste heat from nuclear power stations, and with prawns and flatfish.

The higher temperatures have physiological effects on the maturation of the fish, which may be hindered by the foreshortening of the maturation period; the spawning of pike, ide and perch are hindered by high temperatures and low water levels. Feeding of fish may be adversely affected; there are more empty stomachs and a lowered growth rate in the warmer water (the cultured fish are given supplementary food). Young bream have less of the nutritious copepods and cladocera at temperatures at and above 25°C, and less nutritious detritus comes to dominate the food. Roach feed on zooplankton and chironomid larvae in the cooler water, but on macrophytes and phytoplankton in the heated water.

It was also found that deaths of fish from cold-loving parasites declined in the water above 20°C, but there was an increase of infection by the warm water parasites *Diagramma interrupta*, *Ligula intestinalis* and *Ichthyophthirius multifilis*.

In sum, in this Moscow reservoir temperatures up to 25°C had no adverse effect on fish, but temperatures of 25° to 30°C are in general unfavourable and limiting for fish and invertebrates.

The absence of ice cover where warm water collects attracts many waterfowl in winter, as also do warm springs. Many waterfowl crowded in a small area of open water will cause heavy parasite infection in fish also attracted to the open water.

Experience in Britain and the USA[15] suggests that phytoplankton, unlike zooplankton, are not harmed by passage through condensers. The rooted plants *Potomogeton perfoliatus, Spartina alterniflora* and *Phragmites communis* could tolerate temperatures of up to 35°C. In January, growths of *Cladophora* were found below the outfall but none upstream, on the River Witham in Lincolnshire.[13]

Long-term studies in England have shown no large effects of heat on the composition of the invertebrate fauna of rivers and streams. Possibly some flatworms, leeches and shrimps, and also some molluscs are less abundant below some heated water outfalls; one planarian, *Dugesia tigrina*, may be restricted by warm water discharges.

Some life histories may be accelerated by warmed water. The breeding activities of the isopod *Asellus aquaticus* and the freshwater mussel *Anodonta anatina* were advanced by at least one month in warm water.

In most cases fish are able to avoid uncomfortably high temperatures; in fact, they are attracted to warm water up to 25°C. Because of this, improved fishing may be found during certain months of the year in the warmed water; in the Potomac River, for nine months of the year more fish per trip were caught in the heated than in the unheated parts of the river. But during the hot season, temperatures of 35° to 37°C reduce the numbers of fish and also the species diversity in American rivers.

Little is known about the effect of heated effluents on the migrations of salmon, both European and American. But trout streams, trout and salmon lakes, the headwaters of salmon streams and the cooler hypolimnia of lakes which have a population of salmonids should not be exposed to heated water. Roach and bream had an extended growing season in England, and salmonids reared in warmed water grow faster than those in cold water.

The feeding of fish increases with increase of temperature up to a point. Fish may feed all the year round in warmed water, where they normally cease to feed in winter. But they seem to ingest more vegetable material and therefore have a slower rate of growth. In Ukrainian rivers, feeding intensity of carp was reduced at 29° to 30°C.

Warmed water may advance the breeding season and development rate of fishes. Bleak, tench and pike perch (Walleye) in Poland, and roach in England, all had their spawning season advanced by several weeks in heated effluents.

Heated Effluents in Estuaries

Estuaries have become a favoured site for the building of power stations, which discharge a very large volume of water heated to 10°C above the ambient. This has been feared as possibly adverse to plant and animal life. But at least some of the effects of the warmed water are beneficial. For example, the major power station at Fawley, on Southampton Water in the south of England, warms the water locally in which the exotic American clam or Quahog, *Venus mercenaria*, has established itself, and quickens its rate of growth; there, it breeds earlier in the season than those outside the influence of the warmth.[15] The breeding season of the immigrant barnacle *Elminius modestus* is also extended in the warmed water.

The effect of the heated water from the nuclear power station at Bradwell, on the estuary of the rivers Crouch and Blackwater, was shown to be harmless to the fauna of these estuaries, including the valuable oyster fisheries.

But the heated effluent from the cooling towers of the Hunterston nuclear power station on the Firth of Clyde did have some effect on at least two of the estuarine animals. The intertidal amphipod *Urothoe brevicornis*, which lives in the warmed water, begins to breed two months earlier in the year than those living outside its influence. The resulting longer growing period for the newly hatched young gives them an advantage in size over those not so favoured. Also, the young of the intertidal clam *Tellina tenuis*, a very common species, grew faster in the warmed water. This was to be expected; the enhanced growth rates and extended growing period are being exploited on an experimental scale in the artificial cultivation of flatfish and prawns.

In the warmed water from the Hunterston power station, the water snail *Nassarius reticulatus* is affected in two ways. Directly, the snails affected by the warm water grow thinner shells than those in normal seawater; the same effect has been found in a related species affected by a power station at Long Island, New Jersey. This thinner shell benefits their predators. There is also an indirect effect; for the warm water causes the breeding season of *Nassarius* to be advanced to January and February, instead of the normal April and May. But the newly hatched young stages, or veliger larvae, need abundant phytoplankton as food; those hatched in the unusual months of January and February may be too early to benefit from the normal spring outgrowth of the phytoplankton. The breeding cycle of these organisms is synchronised with that of their food,[16] so that an advanced breeding season may be of no benefit.

In temperate climates, then, it seems that the effect of heated water in the estuaries has been beneficial rather than the reverse, since it has the impact of an earlier and longer-lasting summer growth period. But the situation is different in hotter climates. On the southeasterly coasts of the United States and in the Gulf of Mexico, summer temperatures are already near the upper limit of tolerance for much marine life. The discharge into these estuaries of large quantities of hot effluent could raise the temperature to 38° to 40°C, a lethal temperature for most organisms. It is estimated that by 1980 American thermoelectric power stations will discharge about 80×10^{10} gallons of water per day, heated to 6° to 7°C above the ambient.[17]

Further, large numbers of small invertebrates are sucked into the cooling systems of power stations, and there subjected to severe buffeting, heat and to the weak chlorine concentration added to the water to inhibit the settling in the pipes of sedentary organisms. It was found that organisms circulating in the cooling water system at Hunterston would not be subject to temperatures greater than 24.4°C, and that for only eight minutes. Such treatment is not likely to effect most marine larvae, nor is the low level of chlorine. Turbulence in the pipes must be very violent, but it has been shown that phytoplankton passes unharmed through them.

But a major station is like a giant filter feeder, which, in a very few days, can filter a volume of water equal to that of the entire contents of a large estuary. In American estuaries, this could include twenty million young of the valuable fish, striped bass.

Pollution by General Rubbish

The wholesale dumping of general rubbish in water bodies is one of the uglier features of the modern environment. It is made worse by the great mobility of societies today, which enables people to visit even remote beauty spots in large numbers. Where public spirit is wanting, they will leave their rubbish behind them. As an example from the literature: the popularity of the snowmobile, and the spreading network of roads in Canada has brought intensive sport fishing to even remote lakes. On one small lake, only 290 hectares in area, sixty fishermen were found on the ice cover, using 129 fishing lines. Over 300 trout lay dead on the ice before the afternoon was half over. Important quantities of rubbish, and especially of glass and plastic containers, metal cans and paper and plastic wrappings, are deposited on the bottom of such lakes.

Dank and foul, dank and foul,
By the smoky town in its murky cowl;
Foul and dank, foul and dank,
By wharf and sewer and slimy bank;
Who dare sport with the sin-defiled?

Charles Kingsley. *The Water Babies.*

NOTES

1. Dietz, Robert S. 1972. In *Exploring the Ocean World.* Revised Edn. New York, Thos. Y. Crowell Co., 1972.
2. Bennett, G.W. 1970. *Management of lakes and ponds.* 2nd Edn. New York, Van Nostrand Co., 1972.
3. Howell, B.R. and Shelton, R.G. 1970. 'Effect of China Clay on bottom fauna.' *J. Mar. Biol. Assn.* U.K. (1970) 50: 593-607.
4. Debenham, F. 1952. *Study of an African Swamp.* London, HMSO. 1952.
5. Fort, R.S. and Brayshaw, J.D. 1961. *Fishery Management.* London, Faber & Faber, 1961.
6. Hickling, C.F. 1971. *Fish Culture.* 2nd Edn. London, Faber & Faber, 1971.
7. Mann, K.H. 1972. 'Case-History – the River Thames.' In *River Ecology and Man.* New York & London Academic Press, 1972. pp. 215-32.
8. Cole, H.A. 1974. 'Marine Pollution and the United Kingdom Fisheries.' In *Sea Fisheries Research.* Ed. F.R. Harden Jones. London, Elek Science, 1974. pp. 277-303.
9. Soong, M.K. and Merican, A.B.O. 1958. 'The use of Endrin as a piscicide in tin-mining pools and fishponds in Malaya.' *Indo-Pacific Fish. Coun.* 8th Session, Colombo, 1958.
10. Wolfgang, Jahn v. 1972. 'Ökologische Untersuchungen an Tümpeln.' *Arch. Hydrobiol.* 70 (4): 442-83.

11. Nelson, D.J., Kaye, S.V. and Booth, R.S. 1972. 'Radionuclides in River Systems.' In *River Ecology and Man.* New York & London, Academic Press, 1972. pp. 367-87.
12. Preston, A. 1974. 'Radioactivity in the Marine Environment.' In *Sea Fisheries Research*. Ed. F.R. Harden Jones. London, Elek Science. pp. 305-29.
13. Landford, T.E. 1972. 'A comparative assessment of thermal effects in some British and North American Rivers.' In *River Ecology and Man*. New York & London, Academic Press, 1972, pp. 319-51.
14. Zhitenjowa, T.S. and Nikanorow, J.I. 1972. 'Der Einfluss des vom Konakowschen Wärmekraftwerk abfliessenden Warmwassers auf die biologischen Prozesse im Iwanikowo-Stausee.' *Verh. Internat. Verein. Limnol.* 18: 833-6.
15. Barnett, P.R.O. 1971. 'Some changes in intertidal sand communities due to thermal pollution.' *Proc. R. Soc. B.* 177: 353-64.
16. Barnett, P.R.O. 1972. 'Effects of warm water effluent from power stations on marine life.' *Proc. R. Soc. B.* 180: 497-509. (1972.)
17. Nugent, R. 1970. 'Elevated temperatures and electric power plants.' *Trans. Amer. Fisheries Soc.* 99 (4): 848-9.

Chapter 9 THE TRANSITION FROM FRESH TO SALT WATER

We distinguish between freshwater and seawater organisms, and the distinction is a real one. Over ninety-seven per cent of the world's stock of water is salt, and contains about thirty to thirty-five parts of salt per mille. The remaining three per cent is freshwater, and contains less than one-tenth of one per mille of salt. Water of salt composition between these two categories is rare in nature, and comprises only a very small percentage of the total. So aquatic organisms are adapted to live in fresh or in salt water, and only a small number can live in either or both indifferently. They are called euryhaline.

Adjustment to Changing Salinity

Most animal membranes are semipermeable; that is, they will allow free passage of water but not to many substances dissolved in the water. Thus, if a semipermeable membrane separates two solutions, one more concentrated than the other, water will diffuse through the membrane from the weaker to the stronger solution until both are of the same strength. This diffusion is called osmosis, and the pressure it sets up across the membrane is called the osmotic pressure; this can be measured with an manometer, and it can be high.

As an example, a human red bloodcell contains a weak salt solution (about nine parts per thousand salt) and is bounded by a semipermeable membrane. When a red blood cell is placed in a strong salt solution, such as seawater, water will pass out through the semipermeable membrane into the stronger seawater and the cell will collapse. But if a red blood cell is placed in tapwater, an extremely weak salt solution, water will diffuse into the cell and will burst it.

Life evolved in the sea; so, as might be expected, most marine creatures have their body fluids of about the composition of seawater, and are thus in osmotic equilibrium with the medium in which they live. But this is not always so in the more highly evolved marine animals.[1] For example, the primitive Elasmobranch fishes and the equally primitive Coelacanth have a body fluid slightly more concentrated than seawater, so there is always an inflow of water into the body, which has to be countered. This is done by the fish retaining urea, a byproduct of metabolism, in its blood, instead of excreting it out through the kidneys as urine. So the blood is kept isotonic with the seawater.

Marine teleosts, which are probably descended from fish adapted to life in freshwater but which have returned to colonise the sea, have body fluids weaker in dissolved salts than seawater, so their problem is to check the consequent outflow of water. This they do by drinking the seawater and excreting the salt through special glands on the gills.

Freshwater teleosts have blood stronger osmotically than the water in which they live; so they have to counter an inflow of water by getting rid of the surplus. They excrete a large volume of urine with a very low salt content; and replace the salt lost by absorbing through their gills the traces of salt present in freshwater.

Euryhaline and Stenohaline Fish

The eel, a truly euryhaline fish able to live in either fresh or salt water has been studied. When an eel is transferred from fresh to salt water, there is at first a loss of water to the extent of 4.5 per cent of the body weight; conversely, an eel transferred from salt to fresh water will rapidly gain in weight as water is absorbed. But these processes are not purely automatic; for an active process of adjustment soon makes the fish isotonic with its environment, but weighing more in freshwater than in seawater. The body fluids of the eel in freshwater are more dilute than those of an eel in seawater.

But most fishes are unable to make the quick adjustment possible to the euryhaline eel; thus a freshwater fish placed in seawater, or a seawater fish placed in freshwater, must die. The practical limit seems to be about ten parts of salt per mille; few freshwater fish will survive for long in water more saline than this. But even this degree of tolerance is of great economic value; for many valuable fish will live and even breed in water too salt for irrigation; thus such water will raise a crop of fish even when it is too salt for most land crops.[2] But when a habitat changes from fresh to salt water, most of the fish will die out; and *vice versa*.

There can be catastrophic fish kills; a case which comes to mind is the very high tide of February, 1953, which, among many much more serious disasters, burst the seawalls of the Norfolk Broads, flooding these inland waters with seawater; there were massive and almost total fish kills.

Euryhaline Fish

Genuine euryhaline fish include the grey mullets, the stickleback, flounder, the milkfish of the Indo-Pacific region and the various species of shads. Young grey mullets, placed direct from seawater into tap-water, showed severe shock, quivering and arching the body, with violent trembling of the fins; death often followed. But they could be transferred from seawater to a solution as weak as five parts per mille without distress, and thence, in one more change, to freshwater.[3] In nature, these and other euryhaline fish can take their own time over acclimatisation to seawater, or to freshwater. Most of the commercially very valuable Penaeid prawns are euryhaline, spending their younger days in fresh or brackish waters, and returning to the sea to breed. The giant freshwater prawn, *Macrobrachium rosenbergi*, must still return to the estuaries to breed.

Osmoregulation of Migrant Fish

Fish which regularly migrate from fresh to salt water and the reverse must obviously have good powers of osmoregulation. The eel, which migrates from freshwater to the deep ocean to breed has already been mentioned; less well known is the flounder, which lives far up estuaries and well into freshwater, but must return to the sea to breed. The migrations of the salmon and sturgeon are well known.

The many species of shads are euryhaline. Only two enter English estuaries, the Allis and Twaite shads, and these are now rare. But the American menhaden, which enters coastal and estuarine waters, supports the greatest of all the American fisheries; and the alewife, another shad, has established a breeding population in the freshwaters of the American Great Lakes. The Indian *Hilsa* runs hundreds of miles up the rivers from the ocean on its spawning migration, and is the occasion for important fisheries; in the West African estuaries the local shad is the bonga, *Ethmalosa fimbriata*. In making these migrations, these fish must be able to alter the composition of their body fluids; and their young must also be able to do so on their return migrations.

In the Baltic Sea, there is a complete gradation from seawater at the western end to freshwater in the northeastern Gulf of Bothnia. In fact, the Baltic can be regarded as an estuary. In the marine western waters live the familiar cod, plaice and herring. In the Gulf of Bothnia live the pike, perch, ruff and carp. Races of herrings exist which can live in very weak seawater.

Lake Qarun, United Arab Republic

Over the decades, the attractive Lake Qarun in the western desert, some 80 km. southwest of Cairo, has become more and more saline. When I visited it in 1946, the desert road was still lined with thousands of oil drums, rolls of barbed wire, and all the jetsam of the late war. The sudden appearance of bright green irrigated land, after the miles of dusty yellow desert, was very agreeable to see. Lake Qarun is the relic of the once much larger Lake Maeotis; in historical times it was used by the Pharoahs as a diversion reservoir for the Nile floods. It had a rich fauna of the freshwater fishes so faithfully shown on the tomb paintings and papyri of ancient Egypt.

As recently as the 1930s, the fauna of the lake was dominated by this Nile fauna, including the commercially important *Tilapia zillii* and *T. nilotica*, seconded by the predatory Nile perch, *Lates niloticus*, and by species of the lipped carp *Labeo* spp. The elephant snout fish, *Mormyrus*, was an important part of the catch of the lake; and even the freshwater eel, after its long migration across the Atlantic and up the Nile, contributed 2,400 kg. annually of well-priced fish.

Increase of Salinity

At the present time, Lake Qarun has an area of about 20,235 ha., and a maximum depth of six metres. It lies forty-five metres below sea level, and is the sump into which drain the irrigation channels for the whole of the Fayyum Oasis. The rate of evaporation is very high, so that the salinity has been steadily increasing.[4] In 1920, it was 11.6 parts per thousand, in 1928 it was 17.2, in 1932, 23.4; in 1963 the salinity ranged from 17.2 to 29.4 according to the position of the observing station in relation to the main drainage canals. Thus a gradient of salinity had become established in the lake, varying from slightly saline water, to water of salinity near that of the sea.

As the salinity increased from year to year, the catches of Nile freshwater fish declined. *Tilapia nilotica*, which is not adaptable to saline water, was the first to vanish, though the *T. zillii* continued to flourish; this very euryhaline fish has been caught in the Red Sea at a salinity of over forty parts per mille. The Nile perch and the labeos vanished, and indeed the fishery of the lake was in a very depressed state. As some remedy for this decline, the euryhaline and commercially acceptable grey mullets were transplanted to Lake Qarun each year from 1928. The young are caught in hundreds of thousands in the pumping station at Mex, on the coast of Egypt, as they swim up from their birthplaces in the sea against the freshwater current coming from the pumps. They were flown down to Qarun by the R.A.F. bombers of the day; and photographs show proud fisheries staff alongside the struts and wire bracing of the biplane bomber, while the cans of live grey mullet fry are displayed on the sand at their feet.

Transplantations of Marine Fish

These transplants were a success from the start. They showed themselves, when grown, to be a mixture of several species of grey mullet, and a fine fishery soon developed. By 1935, 341 tons of grey mullet were landed from Lake Qarun, and provided over eighteen per cent of the catch. Then it was found that one of the species, the jumping grey mullet, *Mugil saliens*, was breeding successfully in the lake, and the stock of that species is fully self-supporting. Transplantation of the other species continues.

But a bolder venture was tried in 1935. The common or dover sole, *Solea vulgaris*, a wholly marine fish, which is caught on the Egyptian coast, was introduced into Lake Qarun. The introductions were tentative, and success did not seem assured. The first introduction was of only fifty to seventy specimens of 15 to 20 cm. length. In each successive year, a few more were flown down, until by 1938 about 1,600 fish mainly small specimens of 7 to 10 cm., had been transplanted. But in 1939, all doubts were removed; for in that year a set net for *Tilapia* in Lake Qarun came up with a catch of soles, of a total weight of 38 kg. A commercial fishery for such well-grown and high-priced fish began at once. Soon, it was also plain that the fish were breeding successfully in the more saline parts of the lake. Good feeding conditions were found in the large populations of the shellfish *Abra* and the common cockle *Cardium*, which must itself have been carried somehow into the lake and established itself. Plentiful 'shrimp' of the genus *Gammarus* are also abundant on the lake bottom. In such favourable conditions, the soles flourished, and by 1962 provided the fishery with 900 tons, or over 37 per cent of the total catch of fish in the lake. These Qarun soles fetch prices as good as those from the Mediterranean coastal fisheries, though said to be not so plump.

It will be seen that the bulk of the fish now produced in this lake are of marine derivation through artificial transplantation; and most of the remainder are the euryhaline native fish, *Tilapia zillii*.

Now that the amount of irrigation water available has been so much increased by the High Dam, it will be interesting to see whether the lake once more becomes fresher, and if so, whether the marine fish will die out and the original Nile freshwater fauna return. The recent history of the lake is a notable example of how a stocking programme can take advantage of the changing salinity and fauna of a body of water.

The Salton Sea of California

This is an inland salt lake in the desert country of southeast California, which occupies the Salton Sink, a depression 274 feet (83 metres) below mean sea level. It is separated from the ocean, at the northern end of the Gulf of California, by the delta of the Colorado River, which is very wide and flat, having a gradient of less than two feet per mile (about thirty-eight cm./km.). In the past, this great river flowed to the sea sometimes along the eastern and sometimes along the western side of its delta; on many occasions also it has flowed north into the Salton Sink, to make a more or less transient lake. For example, in 1891 the Colorado broke into the Salton Sink to make a freshwater lake of about 100,000 acres (40,490 ha.). This lake evaporated, leaving deposits of salts which were later mined commercially. By 1901, a very large acreage of the Salton Sink had been reclaimed for irrigated crops, by training the Colorado River along irrigation canals.

In the winter of 1904-5, these works were breached by a great flood of the Colorado and its tributary, the Gila River. Water poured into the Salton Sink, and by the time the river works were repaired, in 1907, a freshwater lake of about 410 square miles (105,130 ha.) had been formed, with a maximum depth of 83 feet (28 metres).[5]

Since 1907, the flow of the Colorado has been under control; but the rate of evaporation is high in the desert, at about six feet (1,850 mm.) per annum; for many years evaporation from the Salton Sink exceeded inflow from the irrigation works; and as the level of the lake fell, its salt content increased. By 1956, the area had decreased to 340 square miles (87,180 ha.) and the maximum depth to forty feet (fourteen metres). The salt content, which was 10.01 parts per thousand in 1910, had risen to 33 by 1956. This is close to the salinity of the oceans; but the proportions of the salts present are different. The water of the Salton Sea, which is derived from the Colorado River, contains relatively less magnesium, potassium and chlorine than the ocean, about the same amount of sodium, and more sulphate, calcium, bicarbonates and carbonates.

The recent history of the Salton Sea much resembles that of Lake Qarun, with an initial freshwater fish fauna disappearing as the salinity increased, and an artificial replacement of these by fish of marine origin. In the Salton Sea, the freshwater fauna of carp, rainbow trout and the euryhaline striped grey mullet, *Mugil cephalus* L., with many smaller species of fish, soon died out. The grey mullet has died out because, unlike the jumping grey mullet in Lake Qarun, it has never bred in the Salton Sea. In earlier days, this fish, which entered from the Colorado River, had provided a limited sport and commercial fishery, which reached its peak in 1918 with 91,000 lb. (42,000 kg.).

To establish or restore a sport fishery in the Sea, many species have been transplanted from 1929 to the present day, chiefly from the Gulf of California. Only eight species of all these became established, and of these the vanishing striped grey mullet, and the shortfin corvina, *Cynoscion parvipinnis* (Ayres), which has also never bred in the lake, should be excluded. The mosquito fish, *Gambusia*, is of no fishery importance, nor is the desert pupfish, *Cyprinoden macularis* (Baird and Giraud), which was, and remains, the only native fish.

The most important of the established transplants are the threadfin shad, *Dorosoma petenense* (Gunther), which feeds on zooplankton and is a valuable forage fish, the bairdiella, *Bairdiella icistius* (Jordan and Gilbert) also chiefly a forage fish, and especially important in the food chain; and the sargo, *Anisotremus davidsoni* (Steidnacher), which is both a game fish and a forage fish for the orangemouth corvina, *Cynoscion xanthulus* (Jordan and Gilbert), the chief game fish, and one which can grow to a weight of 32 lb. (15 kg.).

The Salton Sea is productive water, with a rate of primary production of about 0.75 gm.C/m^2/day, a rate four times as high as, for example, the surface water in Long Island Sound. Organic matter collects on the bottom, together with a mat of blue-green algae. Very luckily, the pile worm, *Neanthes succinea* (Frey and Leukart), was introduced into the Salton Sea accidentally in 1930, and has flourished. This worm burrows into, and feeds upon, this benthic organic matter; the worm is the chief food of the bairdiella, and this fish is the chief food of the orangemouth corvina, of which 346,000 specimens are, or have been, caught annually in the Salton Sea.

But the future of the fishery in the Salton Sea is dubious; for if the salinity rises to, or surpasses, about forty parts per mille, the breeding of both the bairdiella and the sargo will be in jeopardy,[6] and the food chain may break down. The Salton Sea could then become a brine lake, like the Great Salt Lake of Utah.

The Zuyder Zee

The long history of the Dutch struggle against the encroaching sea was notably advanced by the completion, in 1930, of the dam across the mouth of the Zuyder Zee. Much of the area enclosed was empoldered to make large new areas for agriculture, with new towns and villages. The Zuyder Zee had an area of 3,500 km^2, and about 3,000 km^2 remain under water. But with the closing of the dam, in which there are sluice gates and locks, the seawater was shut out, and the water within the dam has been flushed out by water derived from the Rhine to create a 3,000 km^2 freshwater Isselmeer.

Some valuable local fisheries were lost with the change to a freshwater environment, and many fishermen had to be compensated or retrained for other livelihoods. An interesting anchovy fishery and a herring fishery were abolished, and small inshore trawling for flatfish. But an important eel fishery has developed along the dykes. All the mature silver eels from the Dutch inland waters, and also probably many from the Rhine, seek their way to the sea, probing the barriers of the dykes; long fences of netting are laid out at right angles to the dykes, to steer the eels into nonreturn traps placed at the outer ends. The new Isselmeer is shallow,[7] with a depth of only five to eight metres and its salinity gradually fell between 1932 and 1937 from 5.6 to 0.16 parts per mille. These new conditions suited the freshwater fauna of the Dutch lakes and rivers, which now colonised the new Lake Isselmeer. The marine fauna of foraminifera, nematode worms, polychaete worms and small molluscs, which formerly purified the sewage of Amsterdam city, changed quickly to a freshwater group of worms and nematodes which dealt with the sewage equally effectively, converting its nutrients into organisms which are of benefit to the life of the lake. In fact, the freshwater Isselmeer, in spite of turbidity due to suspended silt, which limits photosynthesis to the upper one metre, has a higher rate of basic production at 170 gm.C/m^2/year than the former Zuyder Zee.

The freshwater fauna of ruff, perch, pike and bream provide a fishery which yields at the rate of 20,000 metric tons annually, whereas the marine fisheries of the former Zuyder Zee yielded 12,000 metric tons.

In 1935, during the transition from the marine to the freshwater environment, the adults of a chironomid midge of the *Tendipes* group emerged from the water in enormous clouds, bringing traffic on the roads over the dykes to a standstill. Clouds of these midges on Lake Victoria were mentioned in Chapter 3. The reason for the pest in the case of the Zuyder Zee was that the marine fish which used to control their numbers had been eliminated by the decrease of salinity, while a freshwater fauna able to control them had not yet established itself. Special arrangements were made to encourage the entry into Lake Isselmeer of as many elvers as possible, by suitable openings of the sluice gates; and these elvers both helped to control the midge larvae, and by growing at their expense added to the lake's productivity. At present, the freshwater ruff, *Acerna*, is the fish most effective in controlling these midges.

The Dutch Delta Plan

Since the disastrous flood of February 1953, which broke through the sea dykes and drowned over 1,800 persons in Holland, the great 'Delta Plan' is being implemented, and is closing off by means of massive dykes and sluices the inlets of the Rhine-Maas delta; excepting only the approaches to Rotterdam, the most important Dutch port and one of the world's greatest. The plan is to exclude the sea permanently, in the interest not only of flood control, but of land reclamation. The salinity of these inlets, now cut off from the sea, is changing and, as in the case of the Zuyder Zee, is affecting the fish fauna and the fisheries. The most serious loss is, or will be, one of the most prolific oyster and mussel fisheries in the world; for the oyster beds will now be flushed with freshwater in which oysters and mussels will not be able to flourish. Substantial compensation will no doubt have to be paid to the operators.

Lake Veere

Studies have recently been published on these changes in the new Lake Veere, created by the damming off of the sea, but still with some connection with the Eastern Scheldt by way of locks. Lake Veere has an area of 20 km^2, and soon after the dam was completed and closed, the salinity fell from 16 to 18 parts per mille to 10 to 11. It has not fallen below this, in spite of flushing with freshwater, because of the inward percolation of salt through the dykes. But the effect of lowering the salinity was apparent from the beginning; for, with the cessation of a tidal ebb and flow, all those algae and molluscs adapted to alternate emergence and submergence died off, and the lush growth of brown seaweeds changed to a brackish water algal flora dominated by the green alga *Enteromorpha*, common in brackish situations. The phytoplankton of Lake Veere also changed from one dominated by centric diatoms to one dominated by flagellates.

These flagellates bloomed enormously in the early days of Lake Veere, because the sewage of the town of Middelburg was for many months discharged into the lake; the density of flagellates could reach ten million cells per ml. These algal blooms enriched the mud bottom deposits, and there led to the proliferation of the bottom fauna, including the large crustacean *Corophium*, the worms *Polydora* and *Nereis*, and many lamellibranch clams, all brackish water forms well able to flourish in the changing salinity and the richer feeding conditions. Plaice, in particular, of which a large population was imprisoned within the dykes, found these nutritious foodstuffs on which to cram themselves, and an important fishery exploits these somewhat euryhaline fish, using traps. These plaice have produced ripe eggs and milts, but have not been observed to produce live young; possibly the eggs and young stages are more sensitive to salinity than the grown fish. However in spite of this the fishery continues, with recruitment through the lock gates from the Eastern Scheldt. Two species of fish which breed in the lake, the eel-pout, *Zoarces*, and gobies, are also flourishing.

Plover Cove, Hong Kong

Hong Kong is a city state, with four million people living in a small area. The demand for water has been growing fast. The climate of Hong Kong has a wet summer season, between May and October, corresponding to the southwest monsoon when about 2,170 mm. of rain falls. This must be stored against the dry season, from November to April. In Hong Kong there is no land to spare for reservoirs, so a new reservoir has been reclaimed from the sea. This stores water, not only from the local catchments, but that brought by pipe line from the only stream of any volume in the territory, the humourously-named Indus, and also some of the water bought from mainland China.

Plover Cove was the innermost recess of the large Tolo Harbour, itself so well landlocked that it was and is in use as a typhoon shelter, to which small craft repair on a typhoon warning. Plover Cove was picturesque in a bleak way, surrounded by low hills without much vegetation, and separated from the main Tolo Harbour by a promontory ending in a chain of small islands. It had an average depth of about ten metres, and I saw fishermen at work in their sampans fishing for octopus, and for surface shoaling fish with gill nets. There were some oyster fisheries, and lesser fisheries for the windowpane oyster and mussels, and for prawns. The villages along the sheltered shores each had their paddy fields, irrigated from hill streams, and in these, and in the ponds between them, grew the water hyacinth which has been described in Chapter 3 as such a curse in some African rivers. But here not only was the weed no nuisance, but its growth was encouraged as a useful pig food supplement. Hong Kong is on the edge of the tropics, and has a decided cold winter season, and it is believed that this checks the weed's fecundity.

The building of the reservoir at Plover Cove required a dam to cut it off from the sea, and this was a 'soft-fill' dam, 2,074 metres long across the mouth of the cove, and with shorter dams to fill the gaps between the islands which partly enclosed the cove. The impounded area was some 1,241 ha., and the designed storage capacity was 187 million cubic metres.[8]

When the dam was completed, the water enclosed was, of course, seawater, and this had to be replaced by freshwater. The quality of the water matters little in a dam built, for example, for power generation or for irrigation; but in a reservoir for drinking water quality is paramount, and the engineers' problem was to get rid of the marine flora and fauna, which would die when the environment became fresh, so that it would not pollute the water and induce the outgrowth of troublesome or even harmful plants. These plants are only too well-known to water undertakers; they choke filters and may annoy customers by giving the water earthy, unpleasant flavours. It was important, too, to get rid of all salt; for a salty layer of water close to the reservoir bottom might induce stratification of the kind described on pages 38-9, and lead to foul anaerobic conditions.

Replacement of Seawater by Freshwater

In the cases of the Zuyder Zee and the Delta Plan impoundments, the seawater was flushed out naturally, and over a period of many years, by the waters of the Rhine, which replaced the seawater by freshwater.

But other means had to be used in Plover Cove. Ten powerful electric pumps, mounted on pontoons, pumped out as much seawater as possible before the expected onset of the rainy season. As it happened, the rainy season was delayed and yielded little rain; this was a great help, as pumping could continue until only a small pool of the original seawater was left.

The invertebrate fauna was most abundant at a depth of four feet below ordnance datum, and there consisted of surface dwelling molluscs and prawns, as well as a burrowing fauna. Another region of rich bottom fauna was at sixteen feet below datum, and here consisted entirely of burrowing worms and clams. The richest area of the bottom fauna occurred at twelve feet below datum, with a biomass of 7 $gm./m^2$ reckoned as dry weight; but even on the reservoir bed at twenty-five feet below datum, there was still a biomass of 2 $gm./m^2$, dry weight. As the level of the water fell with the pumping, the fishermen gathered by hand the more valuable invertebrates, such as edible mussels and clams; and the rest of the bottom fauna died as the cove bottom was exposed during the long period of pumping out. Their bodies contributed to the (unwanted) fertility of the reservoir, but less than some other sources.

During the pumping out, the fishermen were actively encouraged to fish the cove to the limit, and their efforts were supplemented by trawling by government research craft. Some thirty-three tons of fish were caught, which works out at the respectable figure of 27 kg./ha. of the whole Plover Cove area.

The reservoir was filled with freshwater from four main sources. These were, the local catchment, which provided fifty-eight per cent; direct rainfall on the reservoir, which provided seventeen per cent; pumped water from the River Indus, a canalised stream draining a sizeable valley in Hong Kong well developed for agriculture, which provided fifteen per cent; and the pipe line from the East River of China, which contributed ten per cent.

Phytoplankton

After filling with freshwater, the bottom mud continued to diffuse out not only sea salt, but nutrient salts derived from the decay of the former fauna. After the first period of flooding, the water from the River Indus contributed most of the phosphate and nitrate. These nutrients nourished the growth of phytoplankton, species chiefly carried in with the Indus water. In the autumn, green algae tended to flourish, followed in winter and spring by the blue-green alga *Microcystis*. These were followed in succession by the diatom *Cyclotella*. The major outgrowths of the phytoplankton followed periods of mixing of the reservoir water by winds. The increases in the abundance of the zooplankton naturally tended to follow those of the phytoplankton, and consisted mostly of copepods and Daphnids, with some rotifers. Preparations were made for coping with the expected 'blooms' of phytoplankton in the first few months, and stocks of the algicide copper sulphate were laid in; but these were not needed.

The Change of Fauna

In 1969, while the freshwater reservoir was still new the primary production was measured at 0.35 gm.C/m^3/day, which can be compared with 0.64 in Lake Victoria and 0.15 in the English Lake Windermere. Such a rate of primary production is of course unwanted in a drinking water reservoir; but in more recent years the producitivty has decreased as the initial fertility has been used up. There has in fact been great satisfaction with the performance of Plover Cover reservoir, so much so, that it has since been enlarged.

In the freshwater reservoir a new invertebrate fauna soon appeared, though a few of the former marine types survived, such as nereid worms and sabellid tube making worms: these are notably euryhaline in any case. The freshwater clam *Limnoperna*, the freshwater prawn *Palaemon* and the amphipod *Allocrangonyx* became abundant. As in the Zuyder Zee, the chironomids swarmed in the bottom mud, to the extent of 20,000 individuals per square metre.

The original fauna of marine fish took longer to die out than had been expected. The gizzard shad, euryhaline in any case, as are all the shads, is very slow in dying out, and has even attempted without success to breed. The grey mullet, almost certainly the striped grey mullet *Mugil cephalus*, was present in the original cove, but more have entered since filling with freshwater. This fish is so tolerant of freshwater that the young are caught in the mouths of streams, and stocked in the freshwater fishponds of Hong Kong. They feed on detritus and aufwuchs, and its infauna such as nematodes, chironomid larvae and copepods. Young grey mullet have an extraordinary ability to gain access to enclosed waters, and when spillways run at Plover Cove, young grey mullet will somehow gain entry; some may also have entered with the Indus water. Their presence in the reservoir would likely be more beneficial than harmful.

The local freshwater fish which entered the reservoir were unimpressive. A predatory wild carp *Hemiculter*, a feeder on insect larvae, zooplankton, tadpoles and fish fry, soon became dominant in the reservoir; but the rest were small species such as the *Gambusia* introduced to control malaria mosquitoes.

Introduction of Fish

At first, with a view to controlling the growth of the phytoplankton, the two plankton eating fish, big head and silver carp, mentioned in Chapter 5 as introductions into the Siberian reservoirs, were introduced as fingerlings. But it was found that, though they grew very fast and reached a weight of 25 kg. in a short time, they did not control the blue blue-green algae. The common carp was introduced to help control the chironomid larvae, for these insects are its principal food.

A surprise was the appearance and establishment of the African cichlid fish *Tilapia mossambica*, which has already been mentioned in earlier Chapters. Some of these fish were introduced to Hong Kong from Thailand in 1954, to test as a possible commercial culture fish.[9] But it did not thrive, and, in particular, was killed by the winter temperatures which occur in shallow fishponds. Its culture was abandoned. But obviously a few must have escaped, and survived where the low winter temperatures of Hong Kong could be avoided; and some of these have come in with the Indus water.

T. mossambica becomes lethargic at about 15°C, and dies after long exposure to 12°C. But, in a large reservoir such as Plover Cove, falling temperatures will take long to affect the whole water body; though a few fish die every cold season, especially following a sudden rapid chill, the species is flourishing and multiplying. As its food consists of phytoplankton, and also aufwuchs, as well as occasional small zooplankton organisms, it should be a useful fish to have in the reservoir, especially since it has been proved that the very acid stomach of some herbivorous *Tilapias* (pH. 1.4) can lyse blue-green algae, and so allow of their digestion.[10] Already, some large specimens have been taken in Plover Cove, and a sport fishery may soon be established. It has no superior as a table fish.

Water Abstraction from Rivers

The abstraction of large quantities of water from a river can have serious consequences; the reduced flow may not suffice to dilute pollutants discharged into it, nor to support the former fish life, nor to keep the river channel clear of silt deposition.[11] Massive water abstraction may allow seawater to encroach up the estuaries and into coastal waters, thus again causing a shift between marine and freshwater organisms.

For example, the Aswan High Dam in Egypt now holds back in Lake Nasser the whole of the flow of the Nile, one of the greatest rivers in the world. From there, it is directed into irrigation schemes, and none now reaches the sea. But the Nile has always been the only river of any size discharging into the eastern Mediterranean, and for millennia has carried its load of silt and dissolved nutrient salts into that barren area of sea.

For some centuries, the flow to the sea has been reduced by the effect of minor barrages; but at all events these did not prevent massive discharges of Nile water into the sea. Such fertility as the eastern Mediterranean had was largely due to this Nile flood, which amounted to about $35 \times 10^9 m^3$ of silt-laden and nutrient-rich water annually.[12] A bloom of phytoplankton followed the Nile discharge into the eastern Mediterranean, and this fed a zooplankton which supported many shoals of fish, especially sardinellas. These provided a fishery, not only off the Egyptian coasts, but north along the coast of Palestine as far as the Lebanon. In these small boat fisheries, the sardinellas were attracted at night by the light of paraffin pressure lamps (a technique imitated effectively on the freshwater sardinellas or sprats of East Africa, as mentioned in Chapters 4 and 5) and were surrounded and caught in pursing nets. But since the annual Nile flood ceased, the catches of sardinellas off the Delta coast of Egypt have fallen from an annual 15,000 tons to a paltry 500 tons.

The load of silt carried down by the river was partly deposited on the coastal banks and on the narrow continental shelf, where it helped to support a fauna of worms and molluscs which fed a wide range of sea fish, including soles. A trawl fishery, based on these bottom fish, has now almost disappeared, and the fishermen are experiencing hard times.

The wide and very shallow lakes in the Nile Delta have been adversely affected. These were among the most fertile bodies of water in the world, with big harvests of grey mullet and prawns.

The River Don, in the Ukraine

Water abstraction from the River Don in south Russia has had serious effects on the fisheries of the Sea of Azov. This was one of the most productive seas in the world, cropping at 80 kg./ha. of fish annually. The Sea of Azov takes not only the discharge of the Don, but also of the Kuban, and its great fertility is due to the nutrients and detritus carried down by these rivers, and to the shallowness of the Sea of Azov, which allows a quick warming up of the water in spring, and a rapid turnover of material.[13] The sea bed consists of great deposits of organic detritus derived from the rivers and from plankton; these deposits are so rich that from time to time there may be catastrophic deoxygenation of the waters, leading to fish mass mortalities. In spring, there develops a huge biomass of plankton.

The Sea of Azov communicates with the far larger Black Sea through the narrow Straits of Kerch, where the salinity may be as high as 11.2

per mille; in the Sea of Azov itself, the salinity diminishes to five per mille in the Gulf of Taganrog, and thence to freshwater in the rivers. Here the fish fauna includes the pike perch, roach, bream, carp and sturgeons. The local fish fauna is joined in summer, when the Sea of Azov warms up rapidly, by many immigrants from the Black Sea, including grey mullets, and the anchovy, clupeonellas and true herrings, which are all plankton feeders. So rich are the feeding conditions that the fisheries of the Sea of Azov contribute ninety per cent of the catches of fish of the entire Black Sea Basin, about 2.75 million cwt. in 1936. It was a very fertile ecosystem, based on a balance between the outflow of the rivers and the inflow of Black Sea water of higher salinity through the Straits of Kerch.

The regulation of the River Don began in 1950, following the completion of the Tsmilyansk Reservoir, and was complete by 1952. The flow of the Don was reduced from twenty-six to nineteen km^3 annually, the balance being diverted to irrigation schemes. The consequences for the Sea of Azov were serious. During impoundment in the reservoir, the River Don loses much of its load of detritus, silt and nutrient salts, thus losing some of its nutrient value. Moreover, the diminished flow of water into the Sea of Azov, allowed the more saline water of the Black Sea to enter, and this more saline water moved up the estuaries of the rivers. The waters of the Gulf of Taganrog have increased in salinity by two parts per thousand, and the cessation of the outflow of low salinity water through the Straits of Kerch no longer attracts the shoals of Black Sea fish. A table shows how great has been the effect of the regulation of the River Don on the Sea of Azov. The table, from Zenkevitvh gives a comparison of some indices used in each of the characteristics quoted.

Table 1

Characteristics	Average before impoundment	Average after impoundment
Dissolved phosphate	2016	650
Dissolved Nitrogen (summer values)	327	58
Biomass of Zooplankton	475	40
Production of Anchovy and *Clupeonella* (centners × 1000)	4990	1844

It has been suggested that a dam might be built across the Straits of Kerch, to control the inflow of Black Sea water; any such dam would include fish passes to allow the continued immigration of the valuable sturgeons and herrings.

Great quantities of aquatic vegetation grow in the brackish Gulf of Taganrog, and this was but little used because a plant eating fish was wanting in the local fauna. So the grass carp, or white amur, already mentioned in the Chapter on artificial lakes, was introduced, as this fish can stand a salinity as high as eight to ten per mille, and is a most efficient devourer of submerged vegetation. It also has a high market value.

NOTES

1. Davson, H. 1959. *A Text Book of General Physiology*. 2nd edn. J.A. Churchill, London.
2. Loya, Y. and Fishelson, L. 1969. 'Ecology of fish breeding in brackish water ponds near the Dead Sea.' *J. Fish. Biol.* 1: 261-78.
3. Hickling, C.F. 1970. 'The Natural History of the English Grey Mullets.' *J. Mar. Biol. Assn. U.K.* (1970) 50: 609-33.
4. El Zarka, S. 1963. 'Acclimatisation of *Solea vulgaris* (L) in Lake Qarun, Egypt.' *J. du Conseil*, 28 (1): 126-36.
5. Carpelan, L.H. 1961. *History of the Salton Sea*, State of California, Dept. of Fish. & Gam. Fish. Bull. No. 113.
6. Lasker, R. Tenaya, R.H. and Chamberlain, L.H. 1972. 'The Response of Salton Sea fish eggs and larvae to salinity stress.' *Calif. Fish & Game* 58 (1): 58-66. 1972.
7. Vaas, K. 1966. 'Lakes in Dutch reclamation schemes.' *Man-made Lakes*, London, Academic Press.
8. La Touche, M.C.D., Smith, A.J.E. and Townsend, M.I. 1970. *Plover Cove Reservoir, Hong Kong*. Rept. of 2nd Congres des Grands Barrages, Toronto, 1970.
9. Annual Repts. of Hong Kong Fisheries Dept., 1954-5.
10. Moriarty, D.J.W. 1973. 'The physiology of digestion of blue-green algae in the Cichlid fish *Tilapia nilotica*.' *J. Zool. London.* 171: 25-39.
11. Gilson, H.C. 1971. 'The impact of water abstraction on river ecology.' *J. Roy. Soc. Arts, London.* Oct. 1971: 753-5.
12. Aleem, A.A. 1972. 'Effect on river outflow management on marine life.' *Mar. Biol.* 15 (3): 200-10.
13. Zenkevitch, L. 1963. *The Biology of the Seas of the USSR*. English translation. Geo. Allen & Unwin, London.

Chapter 10 ESTUARIES

Rivers enter the sea through estuaries, in which the fresh and salt waters mix over a distance of many miles. This ecosystem may be extensive, and include not only the estuary itself, but also transitional areas such as salt marshes, coastal areas where silt from the river is deposited, and those habitats which, though freshwater, show a rise and fall of tide. The upper parts of deltas are important in this regard. The Pearl River Delta of south China is one of the most fertile areas in the world; this is partly due to the rich alluvial soil, and partly to the large rise and fall of tide even well above the salt water. This rise and fall of water level allows of the irrigation of paddy fields, and for the same reason is the site of the greatest aggregation of fishponds in the world.

Though in the typical estuary there is a gradient of salinity such that the salinity decreases from the mouth to the river, in regions where there are great seasonal variations in rainfall, this gradient may be reversed in the dry season. The higher rates of evaporation at the head of the estuary, and the want of water from the river, may result in the salinity increasing towards the head of the estuary.[1] Organisms which breed in estuaries but return to the sea to grow seek an increasing salinity gradient, and the timing of their breeding must have regard to the seasonal rainfall.

Limans

The Russian and Rumanian 'limans' in the northwestern Black Sea are described as estuaries, though extensive sand bars lie across their seaward boundaries. In the spring, the rains and the melt water flow into the limans and then cause a salinity gradient, and also a higher water level, which bursts the sand bars and restores the connection with the sea. But when these floods are over, and the hot weather comes, evaporation in the limans may raise the salinity to very high levels, high enough to be lethal to fish and many other organisms. When the liman is used for the production of fish, a connection between the liman and the sea may have to be kept open artificially, not only to regulate the increase in salinity, but to ensure the stocking of the liman with fish.[2] Limans are large areas of shallow water which warm up quickly and have nutrient rich muddy bottoms abounding in food organisms; so they are good fish producers. In spring, the young of the grey mullets, which have been spawned at sea, approach the coasts of the Black Sea, and are attracted by the flow of fresh water from the limans. They migrate in thousands up the channels into the main body of the limans, where they grow fast throughout the summer. When autumn comes, evaporation of the water in the limans lowers their level to the point where seawater flows down the channel into the liman. The grown mullet and other fish are attracted by this current to return to the sea, and are caught in traps set in the channels. A similar semicultivated fishery exists in the 'valli' of the northern Adriatic, especially around the Venetian lagoon; but there eels, as well as grey mullet, are grown in the shallow lagoons. The autumn fishery in these valli for the migrating silver eels is famous.

Formation of Mudbanks and Marshes

The effects of an estuary may be felt far out to sea. This is obvious where the river is a very large one, for the water and suspended silt of the Ganges and the Amazon are to be seen many miles out. But even a small estuary such as that of the River Tamar in Devonshire has been traced chemically 45 km. out into the English Channel,[3] and would suffice to guide a migrant salmon, for example, to the estuary mouth and thence to the river.

Generally speaking, estuaries are sheltered from heavy seas by bars and middle grounds. These are the result of the transport of soil by the rivers. In the estuary, the coarser particles are deposited first, and contribute to the formation of deltas. The finer particles settle out in the lower part of the estuary, where the admixture of seawater causes the clay particles to flocculate. This process forms fertile estuarine soils, especially where the watershed of the rivers contains good soil. Volcanic soils are especially fertile, and it has been well said that each eruption of the many volcanoes of Java increases the fertility of the estuaries there, and of the fishponds constructed in them.

As silt is deposited, the mudbanks gradually rise until they become exposed at low tides. On this exposed mud, algae such as diatoms and blue-green algae establish themselves, and with their gelatinous sheaths help to bind the surface. Then salt tolerant plants establish themselves, and by binding and entrapping further mud bring the bank above high water level, and turn it into a salt marsh. Such higher plants include the knot grass, *Spartina*, the samphire, and *Ruppia* in temperate climates, and the many species of mangrove in tropical climates.

Some estuaries result from earth movements; such are the *rias* or drowned valleys of, for example, northern Spain and south Devonshire; these also accumulate silt.

The Estuarine Environment

The estuarine environment has been described as 'harsh,' especially because of the rapid changes in temperature, salinity and oxygen content which can occur. The mixing of fresh and salt water is bound to give sudden salinity changes; the rising of the tide over mudbanks chilled by frosts or heated by a hot sun can produce sudden temperature changes; and the presence of beds of green plants and algae on the one hand, and of beds of decaying vegetation on the other, may give sudden changes in the dissolved oxygen content. Only a limited number of organisms can cope permanently with such conditions. Besides these permanent estuarine inhabitants, many marine fish will enter the estuaries on the rising tide to feed, and return to sea as the estuaries drain at low tide.

For many reasons, estuaries are areas of great fertility. Their content of organic detritus is continually renewed from the inflowing rivers, from the death of freshwater organisms killed by seawater, and marine organisms killed by fresher water, and by the mechanism of the so-called 'nutrient trap.'[4] Inorganic nutrient salts entering the estuary are trapped by the growth of plants which incorporate them; plankton carried in is greater in volume than that carried out; and both plankton and detritus carried into the estuary are filtered off by many estuarine organisms such as worms and especially filter feeding molluscs.[5] After digestion, the faeces of these filter feeders, often bound by mucus, further build up the organic content of the estuarine soil. The worms and molluscs are food for many species of fish, and their faeces feed the growth of benthic algae which in turn feed fish such as the grey mullets and many species of prawn.

In temperate climates, the annual leaf fall contributes great quantities of organic matter to the rivers and thence to the estuaries; and in the tropics the mangroves contribute and also the masses of plant material torn from the river banks. In all estuaries, salt marsh plants and beds of reeds and *Spartina* break down to give detritus.

Detritus

This detritus, derived from the breakdown of plant material, may be further enriched through the action of bacteria. A study of an American estuary showed that particles of detritus derived from the breakdown of beds of *Spartina* grass formed ninety per cent of all suspended matter in the water, the growing phytoplankton contributing only the remaining ten per cent. The finer particles of detritus had a consumption of oxygen many times greater than that of the coarser particles; and, further, the smaller particles had a higher nitrogen content than the larger.[6] In fact, the smaller detritus particles might contain twenty-four per cent of protein on an ash free basis, while the larger had only six per cent. This enrichment in protein, which makes the smaller particles good food for the filter feeding organisms, is due to the growth of bacteria on the surface of the detritus particles, and at the expense of some of the carbon they contain. The detritus particles themselves had little nutritive value. Small detritus particles offer a greater surface area for both respiration and bacterial growth than large ones, and so have a higher nitrogen content.

This work was confirmed in the estuary of the River Thames.[7] There the nitrogen content of the mud deposits showed a regular increase with the decrease in size of the particles. This was due to the richer bacterial flora growing on the surface of the smaller particles. The small estuarine snail *Hydrobia ulvae* feeds by sucking and scraping up the estuarine mud, and in a remarkable series of trials it was shown how the particles become enriched in their content of nitrogen.

The fresh faecal pellets of *Hydrobia* have a high carbon content, 9.5 per cent of the dry weight, and a low content of nitrogen, at 0.02 per cent dry weight. These faecal pellets were cultured in suitable conditions, and a large population of nonphotosynthetic organisms developed which was able to synthesise protein by the oxidation of some of the carbon in the pellets. The pellets therefore once again became nutritious food for the snails, and were fed to them. Once again, the nitrogen content fell to a low level. Evidently, the nitrogen in the faecal pellets represents a bacterial population able to fix nitrogen at the expense of some of the carbon.

Algae in the Estuarine Environment

Blue-green algae, at least forty species of which are known to be able to fix atmospheric nitrogen in the presence of organic matter, flourish in estuaries, and may cover the bottom with an algal felt which feeds many fish and prawns directly, and indirectly provides great quantities of detritus. These and other submerged vegetation, such as brown and green seaweeds and other filmentous algae, not only contribute to the detritus, but, like the freshwater aufwuchs, harbour an intense fauna of insects, worms, and crustacea.

Mudbanks become covered with a film of algae, and this may be seen happening in tropical estuaries. As the tide goes down and exposes the mudbanks, these are at first a brown colour, but this colour soon changes to an olive-green or olive-brown. Diatoms which have spent the high water period just below the mud surface migrate to the surface and there proliferate in the hot damp sunshine,[8] greatly adding to the organic content of the estuary.

Detritus as a Source of Potential Energy

All this detritus, whether entering the estuary or originated there, is a source of potential energy for organisms able to use it. As a foodstuff, it improves as it is broken down into ever smaller particles by mastication and digestion and by the mechanical grinding of the shifting soil. In fresh material, this grinding may release enzymes for autolysis or selfdigestion, and there are also the effects of the saprophytic fungi and bacteria. For example, *Spartina* grass is attacked and colonised by the fungus *Sphaerulina pedicellata*, and this and other fungi and bacteria form an important and significant biomass which is fed upon by nematodes, mites and other small invertebrates.

Detritus ultimately breaks down to much smaller particles. Such particles include colloids, and aggregates of large molecules, such as proteins, fats and carbohydrates; and smaller particles such as vitmins, amino-acids, sugars, celluloses, chitin and urea. Even the gases methane, ammonia and H_2S occur in detrital material. All these substances represent potential energy originally derived from photosynthesis.

Much organic matter is carried into estuaries from the watershed of the rivers; the resistant sporangia of ferns and mosses, and the spiral and annular thickening of vascular plants, are common in the stomach contents of grey mullet, fish which feed in part on detritus.[9] This emphasis on detritus in the economy of estuaries is justified by the importance of the detritus cycle and of the benthic flora. Phytoplankton plays only a minor role in the primary production of an estuary. In an impounded area of estuarine water, it was found that the total gross primary production was 445 gm.C/m^2/year, and that only 98 gm. of this was contributed by the phytoplankton, and all the rest by the benthic vegetation, chiefly benthic diatoms. In this area of water, the consumption of the biomass exceeded production, so that the balance of energy needed was got from the detritus present.[10]

Benthic Production

It is recognised that the growth of the phytoplankton is poor in both the Russian limans and the Italian valli, and that most of the high rate of primary production is due to an intense benthic algal flora, in which benthic diatoms and benthic algae such as seaweeds are the chief producers of new organic matter. In estuaries in Florida, the benthic turtle-grass beds produce at the rate of 34 gm.C/m^2/day;[11] and, though they are exposed to great variations of temperature and degree of desiccation seaweeds of all kinds form dense mats of vegetation in which the rate of primary production may be ten times as great as that of the phytoplankton.

The solid matter in sewage effluents is a rich source of detritus; modern conurbations tend to be established on estuaries, which have been dubbed 'septic tanks of Megalopolis.' Such estuaries are very liable to heavy additions of organic matter, which becomes pollution when it exceeds the natural capacity of the waters to purify by chemical and biological means.

Estuarine Fish Farming

Because estuaries are such fertile ecosystems, a fish farming industry has become established in many localities in the civilised lands of the Far East. It is a traditional technique, yet shows many features in common over a geographical area ranging from Java and Sumatra through the Philippines to Taiwan and the Pescadores Islands. The degree of sophistication varies from country to country, yet there seems no doubt that all have a common origin.[12]

The technique is based on the observation that some valuable fish and prawns feed on the algal felt and detritus typical of fertile estuaries, and that the growth of these foods can be encouraged by simple cultivation methods. The culture is efficient because the food chain is very short, namely, from plant (alga) direct to fish and prawn, with the minimum loss of material in the food chain.

The felt of growing algae on firm bottoms and on substrates such as gravel, timber and submerged plants, consists of filamentous green and blue-green algae, with their load of epiphytic diatoms. The presence and importance of these epiphytic diatoms is proved by the ash content of the algal felt; for though the algae have little silica, the ash has a high silica content due to the frustules of the epiphytic diatoms.[13] This algal felt is given some cultivation by tilling the soil when it has been drained dry; this helps to mineralise some of the organic matter in the pond soil. When the soil shows signs of exhaustion, its fertility may be renewed by the application of compost. This usually takes the form of grasses and weeds and mangrove leaves placed in heaps in the ponds to a dosage of about 3,000 to 4,500 kg./ha., repeated two or three times at three-monthly intervals. The effect of this treatment has recently been studied.

The treated ponds differed from the untreated controls in the following respects: they had a lower dissolved oxygen content and a higher CO_2 content at night; but also a seven-fold increase in the abundance of the benthic blue-green algae, a three and a half fold increase in the growth of the filamentous green algae, and a four-fold increase in benthic diatoms. Evidently this traditional treatment is effective.[14]

Management of the Ponds

In Taiwan, where the winter temperatures fall to the point where the growth of fish is much reduced, it is customary to drain the ponds and overwinter the fish stock in a special deep ponds protected from the prevailing northerly winds. While the main rearing ponds are thus dry, they are tilled and fertiliser is dug in. Organic fertilisers such as rice bran, chicken dung or oilcakes are used, and the pond bottom is levelled to give a firm surface for the growth of the algal felt when the ponds are refilled in spring. So effective are these methods of husbandry that, in spite of the check due to the colder season, an eight-month growing season in Taiwan gives the highest fish crops in brackish water recorded, at about one ton per acre annually.[15]

The algal felt consists of several filamentous species, such as *Phormidium* and *Lyngbya* and their associated epiflora of diatoms. The close connection between the detritus (organic) content of the pond soil and the growth of the algal felt is well shown by a comparative study of groups of ponds in Taiwan. Grouping these ponds according to their soils the following relations were found:–

Table 2 From Tang and Chen.[16]

Percentage of Organic Matter	Annual Crop of Algae (kg./ha. wet weight)	Fish kg./ha.
1.23	15,000	1,200
2.41	18.750	1,500
3.27	25,000	2,000
4.17	28,250	2,500

The table shows that the algal crop increases, though not proportionately, with the increase in the percentage or organic matter (presumably measured by the loss on combustion); and that the conversion rate of wet algae to fish is about twelve, a high figure suggesting nutritious food and good digestibility.

The findings of the table are in line with the fact that organic fertilisers are most effective in these brackish water ponds, and it has been found that the inorganic fertilisers, such as nitrates, phosphates and mixtures of these, are ineffective. But these estuarine ponds are shallow, seldom deeper than half a metre, and production is mainly be benthic plants. Urea has shown promise, but this substance comes into the class of organic fertiliser.

The successful modifications of technique in Taiwan are being adopted in the Philippine Islands, which have long had estuarine fish ponds.

Produce of Estuarine Fishponds

The produce of the ponds are fish and prawns; in the tropics, the fish is chiefly the milkfish, *Chanos chanos* Forsk., a herring-like fish which breeds in the open ocean and there grows to a length of about a metre or more. In the ponds they grow fast, but are marketed at a weight of a half to one kilo. This fish is wholly euryhaline, living well in fresh water; and it is also toothless. With milkfish, or in substitution for them in temperate climates, are grown grown the grey mullets, chiefly the striped grey mullet, *Mugil cephalus* L., likewise euryhaline and toothless. Both fish feed on the aufwuchs and its infauna and epifauna, and both take in much silt with their food to act as a grinding paste; both also have very long intestines and a strong muscular 'gizzard,' probably to help force the food along the gut. The prawns are both detritus feeders and carnivores, tha latter feeding on small crustacea.

Neither milkfish, grey mullets or prawns breed in the ponds, which are stocked with young bought from dealers or carried naturally into the ponds. The young fish are caught on the open beaches or in the estuaries and stream mouths by specialist fry fishermen. The young of the prawns, which spawn at sea, drift or swim into the shallow coastal waters and estuaries and are passively swept into the ponds with the inflowing water when the ponds are filled at high tides. Recently, grey mullets have been successfully spawned in captivity with the aid of hormone injections, but the problem of raising an economical proportion of the young to commercial size and weight has not yet been successfully solved.[17]

All these animals feed on the rich algal felt and the detritus on the pond bottoms. As the fish grow, they graze down the algal pastures, and may have to be culled, or transferred to new ponds with an uncropped algal felt. In Taiwan, as the natural algal pasture is grazed thin, pelleted foods may be given to keep the growth of the fish at a fast rate. But there are also serious competitors for the algal pasture. The larvae of chironomid midges may develop in immense numbers on and in the algal felt, and in the summer may consume 60 to 90 kg./ha./day. This quantity is as great as the needs of the growing fish crop. No fishpond, however fertile, could produce enough algae to satisfy both, and so the fish production suffers. The chironomid midges also burrow into and loosen the algae, destroying the productive beds. Various treatments, for example with gamma-BHC, are used to control the larvae, and also other lesser competitors such as Cerithiid snails, and this has resulted in better harvests of fish.

Brackish water fish farming is not a negligible industry; including the extensively cultivated Black Sea limans and Adriatic valli already mentioned, there are probably about 400,000 ha. of ponds, producing about 200,000 tons of high grade fish and prawns.

Estuarine Fisheries

There are considerable fisheries in estuarine conditions; in fact, most of the fish landed in America are caught in conditions more estuarine than oceanic. There are the salmon fisheries, where the ascending fish may be caught in drag nets or in fixed traps, and the smaller seines or tuck nets which capture such fish as the flounder and bass and grey mullet. Prawn trawlers work in the deeper channels, and oyster dredgers use the tidal currents to dredge the oyster beds. In China, the prawn fishing junks in the Pearl River Estuary tow several trawl nets simultaneously from booms, hauling in the nets one after the other, and resetting them, without stopping the vessel. There are important cockle, clam and mussel fisheries in estuaries and in tropical estuaries the valuable mangrove crab, *Scylla serrata*.

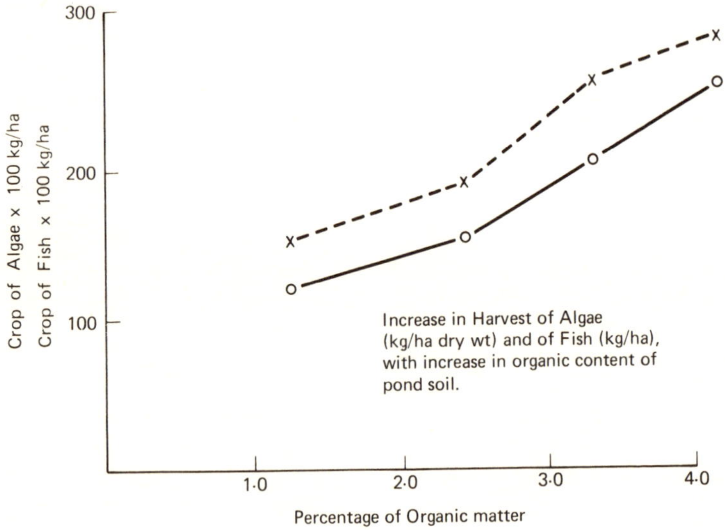

Figure 5 Relation between organic content of soil and rate of fish production in brackish water fishponds[16]

An interesting and important estuarine fishery is that for elvers, the metamorphosing young of the true eels. Best known in Britain is the elver fishery in the estuary of the River Severn based on the hamlet of Epney. In the spring months, on the rising tides, the elvers come swimming up the estuary, keeping close to the banks, especially in the evening. The fishermen use lights to guide them, and scoop the swarms of elvers out of the river with long handled scoop nets. The living elvers may be sold to local dealers to make elver pies, a well-known Gloucestershire delicacy; but there is also a good sale for the live elvers to German and Polish dealers who export them to eastern Germany and Poland, where there are large areas of fresh water too remote from the ocean to receive in the natural way a sufficient supply of elvers. Recently, Japanese buyers have entered this market, since Japanese and Taiwanese estuaries do not give the very large quantities of live elvers needed by the Japanese eel farming industry.

Fixed Fishing Gears

There are many kinds of fixed gears in estuaries besides those for salmon. The biggest of these are the *dōl* nets of the Bombay area, and they are very large indeed. They are anchored in water as deep as sixteen fathoms, and are free to swing with the tidal currents. Each team of fishermen numbers eleven or twelve, and they work two *dōl* nets, one at sea fishing and one ashore for care and maintenance. Large catches of prawns and small mysids of the genus *Acetes* are made, and plenty of fish with them, including some big sciaenid predators and the *Harpodon*, a compressed fish with long teeth which is dried to make 'Bombay Duck' served with curries. This *dōl* net works on the simple principle that the prawns and fish are swept more or less helplessly by the strong tidal currents into the long net. A similar but very much smaller net, the 'stownet,' is used in the Thames estuary. The fishing boat uses an anchor heavy enough to hold both itself and a long net supported from poles and lowered beneath the hull of the boat. Sprats, whitebait, shrimps and prawns, small octopi and other estuarine organisms are carried by the current into the net, which tapers to a pocket from which the catch can be removed at intervals.

Prawn traps of many kinds are set in tropical estuaries. They are variations on long tapering nets set on stakes driven deep into the mud. They fish usually on the flood tides, and are set in rows, one row behind another, until the estuary seems full of them. Each fisherman will tend several. When the tide turns, and the nets are not fishing, they are lifted from the water and draped over stakes to dry.

In the Malaysian estuaries massive fish and prawn traps abound; and a description of a *jermal* will suffice for them all. A platform is built on a supporting framework of strong piles driven deep into the seabed, and may be also supported by guys against the force of the tidal currents. From this platform two long lines of stakes run out in the form of a 'V' facing the direction of the falling tide. These lines of stakes, several hundred yards long, sway and vibrate in the strong water flow, and guide the prawns and fish between them to the platform, from which a long filter net, mounted on long poles operated by turnbuckles, now streams out. The fish buyers' sampans moor to the platform, on which there is often a thatched hut, and buy the prawns and fish on the spot. A popular item of the catches is the 'ikan bilis,' the small anchovy *Stolephorus indicus*, which is dried for use as a relish or as a side dish with curries.

Fish Traps

The changes in water levels caused by the tides in estuaries are much used for fish traps. The simplest fish weirs are of stakes or brushwood, or even of stones, constructed in such a way that the fish and prawns are guided into nonreturn chambers from which they cannot escape, and in which they can be picked up at low water. All kinds of estuarine fish are to be caught in these weirs, sprats, flounders, grey mullets, bass and herrings. They may be more elaborate, made of netting supported on poles, and of varied complexity of design; there may be several guiding nets and traps.[18] A portable form of this net is the stopnet, one of the universal fishing devices. It is a long and deep net which is spread at low tide in a semicircle concave to the shore at places known to be visited by fish on the rising tide. The footrope is trodden into the mud so that, as the tide rises, there is no hindrance to the fish passing over the net. But at or near high water, the headrope of the net is raised and mounted on long poles. The fish are thus trapped as the water falls, and can be picked up. A crew of several men is needed, but the method is a simple and successful one.

Effect of Severe Winters

Shallow estuaries, a 'harsh' environment, are liable to sudden changes of temperature, and a study has been made of the effects on the fauna and flora of estuaries of the severe winter of 1962-3. In the River Blackwater, on the southeast coast of England, the effect on the animals was two-fold. They were subjected to very low air temperatures when exposed at low tides, and shore crabs, *Carcinus maenas*, were almost all killed, and polychaete worms died in large numbers. The other severe effect was the mechanical damage done by heavy ice floes, up to one metre in thickness, as they stranded on the beach or were moved about over the shallows. Great numbers of burrowing shellfish, such as cockles and *Scrobicularia*, were crushed, and seventy to ninety-five per cent of native oysters, and ninety-seven to one hundred per cent of relaid Portuguese oysters, were killed. But the two chief oyster pests, the whelktingle *Urosalpinx*, and the slipper limpet, *Crepidula*, which are unwelcome immigrants from the eastern seabord of the United states, were little affected, since in their native habitat they regularly endured the conditions they met with in 1962-3. Their mortality rate was less than twenty-five per cent. Cockles not only suffered from the cold and ice damage, but from the predation of seagulls and oyster-catchers, for which they were the only food available.[19] Even so, many of these handsome birds died of starvation. Mussels suffered less direct mortality, but were rendered so torpid by the cold that their shells were easily pried open by predators. Estuarine fish such as the grey mullets and bass were killed, both by stranding and by the crushing of the ice floes. Other fish which suffered mortalities were the conger eel and the ballan wrasse,

These extreme conditions had some long-term effects. For example, newly-settled young mussels were very plentiful in the summer of 1963, both in the estuaries and along much of the English east coast, colonising areas where they had not been seen for years. Their chief predator, the shore crab, had been so reduced in numbers by the cold that the mussel spat had the chance, usually denied them, of settling in extraordinary numbers.

NOTES

1. Day, J. 1967. 'The biology of the Knysna estuary, South Africa.' In *Estuaries* Vol 2, (Lauff G Ed): 397-407. Amer. Assn. Adv. Sci. Washington D.C.
2. Ilin, B. 1954. *Mullet Farming*. Krimizdat. (In Russian.)
3. Armstrong, F.A.J. and Butler, G.I. 1962. 'Hydrographic surveys off Plymouth in 1959 and 1960.' *J. Mar. Biol. Assn. U.K.* (1972) 42: 445-63.
4. Hedgepeth, J.W. 1967. 'The sense of the meeting.' In *Estuaries*, as above: 707-10.
5. Korringa, P. 1967-8. 'Estuarine Fisheries in Europe as affected by Man's multiple activities.' In *Estuaries*, as above:
6. Odum, E.P. and de la Cruz, A. 1967-8. 'Particulate organic detritus in a Georgian salt marsh estuarine ecosystem.' In *Estuaries*, as above: 383-8.
7. Newell, R. 1965. 'The role of detritus in the nutrition of two marine deposit feeders, the Prosobranch *Hydrobia ulvae* and the bivalve *Macoma balthica*.' *Proc. Zoo. Soc. London* 144: 25-45.
8. Harper, M.A. 1969. 'Movement and migration of diatoms on sand grains.' *Br. Phycol. J.* (1969) 4: 97-103.

9. Hickling, C.F. 1970. 'A contribution to the natural history of the English Grey Mullets (Pisces, Mugilidae).' *J. Mar. Biol. Assn. U.K.* (1970) 50: 609-33.
10. Malone, T.C. 1969. 'Primary productivity in a Hawaiian fishpond and its relationship to selected environmental factors.' *Pacif. Sci.* 23: 26-34.
11. Odum, H.T. 1957. 'Primary production measurements in 11 Florida springs and in a marine turtle-grass community.' *Limn. Ocean.* 2: 55-112.
12. Hickling, C.F. 1970. 'Estuarine Fish Farming.' *Adv. Mar. Biol.* 8 (1970): 119-213.
13. Schuster, W.H. 1952. 'Fish Culture in the brackish water ponds of Java.' *Special Pubns. Indo-Pacific Fish. Council* 1: 1-143.
14. Saanin, H. and Tati, Ramelan, 1966. 'Fertilisation and its effects on the hydrology of brackish water ponds.' *Proc. 11th Pacific Sci. Congr. Tokyo*, 1966.
15. Lin, S.Y. 1968. 'Milkfish farming in Taiwan.' *Fish Culture Rept. No. 3*, Taiwan Fish. Res. Inst.
16. Tang, Y.A. and Chen, T.P. 1966. 'A survey of the algal pastures of milkfish ponds in Taiwan.' *FAO World Symposium on warn-water fish culture.* Rome, 1966.
17. Anon. 1972. 'The Grey Mullet: induced breeding and larval rearing.' Oceanic Institute, Waimanalo, Hawaii.
18. Davis F.M. 1936. *An Account of the Fishing Gear of England and Wales.* 3rd edn. Fish. Invest. XV No. 2. London, HMSO 1936.
19. Crisp, D.J. Ed. 1964. 'Effect of the severe winter of 1962-3.' *J. Anim. Ecol.* 33: 165-210.

Chapter 11 THE TIDAL ENVIRONMENT

The environment which lies between the level of high and low tides on the coasts is at least as harsh as the estuarine one. At high tides there is the buffeting of the surf, and the battering of ice floes in cold climates. At low tides, those organisms which cannot retreat with the ebbing water are liable to desiccation, and, if they take refuge in small rock pools, to rapid variations in temperature and salinity. Intertidal organisms may suffer great heat in summer, severe frost in winter. But they can survive and flourish by a variety of stratagems. They may burrow in the sand, and there find not only shelter but a more equable temperature and salinity than on the surface. There is shelter on rocky shores in crevices and among masses of seaweeds, while many animals have flattened or even sucker-like bodies which cling to the rock surface and offer no resistance to surf. Such are the limpets and the chitons, which can close their shells tightly enough to prevent loss of water while exposed at low tides. Acorn barnacles, which cement themselves to their substrates, have a low profile and close their cover of limy plates against desiccation; those barnacles which have settled high up the beach can stand many days of exposure, reopening to feed when the sea again covers them. Mussels attach themselves to solid substrates by a strong web of horny byssus threads, and their shells are streamlined against the tug of strong waves.[1]

Sandy Shores

Sandy shores contain varying amounts of silt, often rich in organic matter, chiefly derived from estuaries, and often carried by the currents far from its origin.

Detritus with a high organic content is a good food for many intertidal organisms, as it also is for estuarine organisms. The well-known lugworm, *Arenicola marina*, feeds on soil rich in organic matter, and in such situations occur by the thousand, as many as 82,000 per acre, or eighteen to the square yard. Their casts of coiled tubes of sand, thinly bound with mucus, mark the position of the tail, and a shallow conical pit near by marks the position of the head; in a U-shaped burrow between these marks lives the worm, easily revealed by a thrust of a spade; a staple and easily caught bait for hand line fishermen. A sea cucumber, *Synapta*, lives in a similar way. Many specimens of mollusc, such as cockles, clams and razorshells, live buried in the sand, but thrust up two long tubes, or siphons, which extend above the surface of the sand when covered with water. The shellfish inhales water down one of these siphons, and passes it over elaborate gills, which have a double function. Like true gills, they exchange dissolved oxygen from the water for the carbon dioxide and other waste products of the animal; but they are also an effective and sensitive feeding apparatus, filtering off the particles suspended in the water, and sorting them out, by means of palps, into edible fragments, alive and dead, which are passed to the mouth, and inert fragments such as grains of sand, which are passed out of the exhalent siphon with the spent water. Just above the surface of the soil there is often a layer of fine pale green detritus floc, rich also in very small plants, animals and bacteria. The siphons of the lamellibranchs tap this layer, which is able to support very large populations of shellfish. The common cockle, *Cardium edule*, is a successful animal, able to live and flourish in brackish water as well as in the very high salinities encountered in some of the Russian limans; it can endure great heat and cold. In favourable conditions, this shellfish can form an almost continuous pavement just beneath the surface of the sandy mud, to the extent of over 1,000 specimens per square metre.[3]

Cockles and Clams

Cockles are not only food for humans. When they are young, they are a preferred food for many fish such as cod and flatfish, skates and rays. Also, at low winter tides, cockle beds are the profitable winter hunting grounds for many wildfowl, including gulls and especially for oyster-catchers. It has been proved that these beautiful birds eat a high proportion of the cockles on the beds, to the detriment of the cockle fishermen. In the years 1961-2 to 1964-5, they devoured 600,000 to 1,300 million two-year old cockles, from a total stock estimated at 1,000-3,000 millions: the fishing catch was much less.[2] At the time of writing, a campaign of shooting the birds is causing a heated public debate.

Relatives of the cockle, such as the clam *Spisula*, occur in millions, and are very important to the sea fisheries. On the Dogger Bank in the North Sea, great areas of the sea bottom are covered with them; such patches may exceed 1,000 square miles in area, and in them the clams may occur to the extent of 1,000 to 8,000 per square metre.[4] All are supported by the detritus originating from dead plankton and from the tidal environment. *Spisula* is a most important item in the diet of plaice and cod, and in the normal way there are not enough plaice to use any but a small part of this well-spread feast. Hence plans made many decades ago, to transplant small overcrowded plaice from the nursery grounds in the shallow waters of the eastern North Sea to the Dogger

Bank, where they grow very fast. But this sensible and practical plan to make better use of nature's bounty has always foundered on the international status of the Dogger Bank, far outside territorial waters of any country.

The well-known razorshells, also good bait for hand lining, can burrow fast when exposed at the surface of the sand, using the protrusible 'foot' to secure some purchase in the sand. The clam *Tellina* also lives buried in the sand, but its siphons are much used as food for fish, and especially flatfish. The clam soon grows another pair; but the material used up in this repair may prevent the breeding of the clam.[5]

The shellfish generally are an underdeveloped resource for mankind, though the favoured oysters may be locally overexploited. Clams of all kinds are edible, but few species are exploited, one reason being that the cost of gathering and preparing them is high. Harvesting machines have been invented for oysters and for some clams such as the Quahog on the east coast of America. Shucking machines are in use to mechanise the removal of the meat from oysters and clams; and the meat of mussels is separated from the shell by passing them under spring loaded rollers which crush the shell without damaging the meat, which can then be floated off the heavier shell.

The Toheroa of New Zealand is such a delicacy that it is protected, and digging for them by picnickers controlled. The Abalone of the Pacific coast of America is another well priced mollusc, formerly plentiful enough to be canned commercially. In the Channel Islands, the local Abalone, the ormer, now has to be protected, since aqualung divers can easily work over the rocks on which the ormers browse.

Filter Feeders

Many worms feed, not by swallowing the sand, but by filtering off the particles of detritus on a circle of ciliated tentacles. Such worms may be present in thousands per acre, and their tubes, protruding just above the level of the sand, mottle the surface. When they are covered with water, the worms extend their crown of tentacles and begin to feed. Each tentacle has a cover of cilia, which beat incessantly and drive the fine particles of detritus towards the centrally placed mouth. These worms are also liable to have their tentacles bitten off by roving soles. Some sea anemones live in the same way, buried in the sand. Though all these sand dwellers are sedentary when adult, they have active planktonic larvae which are usually totally different in appearance from the adults. These free living planktonic larvae guarantee the dispersal of the species.

Burrowing Crustacea

But there are also many active inhabitants of the sandy intertidal environment, such as the predatory crabs and the shrimps and prawns which live by filtering off small planktonic animals and detritus in their thoracic limbs, which are fringed with fine hairs. The crabs can shuffle sand over themselves and await the rising of the tide. In the tropics, the burrows of the ghost crabs are as striking as the worm burrows on temperate beaches. Each crab comes scampering out of its hole at high speed as the tide falls, to run with its fellows in swift groups into and out of the breakers, too fast to catch, their eyes set on long stalks, to watch every move of possible enemies. From their burrows at low tide run long radial lines of faecal pellets, their food being the rich silty sand and anything they can scavenge. Where the beaches are the public latrines, these crabs are important as cleansers. These ghost crabs are the prey of many species of fish, espcially the sciaenids, or croakers.

On sandy beaches there are many semiburrowing higher crustacea, such as amphipods and isopods. Amphipods are very common along the tide lines, where they shelter among the windrowed seaweeds on which they also feed. Stomachs of grey mullet I examined in the Isles of Scilly were well filled with these amphipods, snatched up as the fish browsed among the heaps of disintegrating seaweed.

Sand eels live buried in the sand at low tide, and can be caught by quick flicks of a spade just below the surface of damp sand. The young stages of many fish, especially plaice, dabs and soles, may also spend the low water period on sandy beaches buried in the sand of shallow pools. Yet another interesting sand dweller is the masked crab, *Corystes cassivelaunus* which has adapted to life in sand. It has long and very hairy antennae, which, when pressed together, form a tube down which water may be drawn for respiration and for food filtration, very much like the lamellibranchs already described. They can live buried as deep as seven inches in the sand.

The Interstitial Fauna of Sand

An important group of small animals lives in the spaces between the grains of sand. This interstitial habitat has a more equable climate than the sand surface, for a few centimetres below the surface diurnal changes of temperature are slight. The most important of these interstitial animals are the harpacticoid copepods and the free living nematodes; but there are lesser numbers of turbellarian worms, gastrotrichs and tardigrades or bear-animalcules. The mean density of this biomass may be as high as 394 per 10 cm^2. This fauna may descend, in winter, to a depth of 50 cm. to avoid the cold.[6]

This sandy meiofauna feeds on diatoms and bacteria, which grow on the surfaces of the sand grains. The nematodes scrape this material off the sand grains, and the harpacticoid copepods also feed on green flagellates as well as on the diatoms. In favourable localities, the biomass of the meiofauna may attain 16×10^6 organisms per square metre.[7]

This meiofauna is food for many small fish, and for prawns and shrimps and sand-dwelling hydroids and the predatory bristle worm *Nereis*. Grey mullet, *Crenimugil labrosus*, of all sizes devour large numbers of the nematodes and copepods as they browse over the algal felts and organically rich soils, and these are by far the most important animal items of their food.

On tropical sandy beaches, *Uca* crabs feed on the nematodes, which they find by turning over the soil; they can crush the tough worms in their mouth parts.

On sandy beaches on the north and west of Scotland the worm *Eurydice* was the only commonly-occurring organism in shifting sand, while amphipods were frequent on exposed beaches. On more sheltered beaches there were many polychaetes, though the clam *Tellina tenuis* had the greatest biomass.

Eel Grass and Turtle Grass

The eel grass *Zostera*, a true higher plant with a root system which helps to bind the soil, grows in sheltered areas of muddy sand. It is an important primary producer, especially of nutritious detritus when its tissues break down. Also, its long strap-shaped leaves are a good substrate for epiphytic plants and animals, on which many browsing molluscs feed. It has been shown that most of the fertility of the enclosed waters of the southern Baltic and Belt Sea is due to detritus derived from *Zostera* beds.[8] In the warmer waters of the Mediterranean, *Zostera* may be replaced by *Poseidonia*. So clear is the eastern Mediterranean, with its lack of plankton, that I have seen living *Poseidonia* brought up entangled in trammel nets from a depth of 150 feet off Cyprus. In the tropics, turtle grasses and manatee grasses (*Thalassia* and *Cymodocea*) cover large areas of sand flats inside barrier reefs and atolls. Like the homely *Zostera*, these sea grasses perform a valuable function in fixing loose soil, in having a high rate of primary production and in offering shelter and food to large numbers of small animals. Growing in areas very poor in phytoplankton, these grasses are enclaves of high productivity. On Long Cay in Florida one bed of turtle grass produced 34 $gm.O_2/m^2/day$, equivalent to 11 $gm.C/m^2/day$. In the Laccadive Islands, the seagrass beds in an atoll lagoon had a net rate of production of 5.8 $gm.C/m^2/day$.[9] These beds are well browsed, and their breakdown supports the sparse animal communities of the shallow lagoons.

For example, detritus from turtle grass, *Thalassia testudinum*, was found to harbour, per square metre, 3×10^9 bacteria, 5×10^7 flagellates, 5×10^4 ciliates, and 2×10^7 diatoms.[10] Thus detritus consumed from 0.7 to 1.4 mg.O_2/gm. dry weight. The detritus-feeding amphipod, *Parhyallella whelpleyi*, further comminuted the detritus into even smaller particles, thereby increasing its rate of oxygen consumption by 110 per cent. This detritus would be a nutritious food for many other detritus and filter feeding animals, and these would support a fauna of predatory crustaceans, such as the crawfish common on coral reefs, and fish.

Active Swimmers

At high tide, predators such as fish, squids and cuttlefish swim over the sand. The cephalopods, using their efficient jet propulsion, can spring on prey such as crabs, prawns and small fish, gripping them in their tentacles and killing them by biting with their sharp horny beaks, which are equipped with poison glands. The digestion of the prey external, by digestive juices which dissolve the prey for swallowing by the cephalopod.

Rocky Shores

Seaweeds cannot grow on sandy beaches, except where stones offer a firm hold, and for the same reasons encrusting animals are also absent. But rocky shores support a flora and fauna equal in productivity to most other areas of the earth's surface, and higher than that of the phytoplankton.[11]

The basis of this production is the mass of seaweeds of many species belonging to the brown, red and green groups of these algae, which grow in profusion on the rocks and on each other. Like the freshwater aufwuchs, they are able to flourish because, when covered with water, they have access to continuous and renewed supplies of dissolved carbon dioxide and nutrient salts. They differ in that, depending on their situation, they are exposed for several hours daily without water, and so their rate of production must be much reduced unless they live in rock pools. Like the aufwuchs of freshwaters, they harbour and nourish a great fauna of small animals of every kind.

The Seaweed Crop

The density of the growth of seaweeds must reflect the local fertility of the sea. For example, the shores of the eastern Mediterranean, where the nutrient salts in solution are scanty, have poor weed cover. But on temperate shores, especially where upwelling of deep water brings nutrient salts into the surface water, seaweed growth may be luxuriant. On such coasts as those of California, Peru, Chile and around the Falkland Islands, great beds of the giant kelps *Macrocystis* grow in underwater forests, thick enough to act as a useful breakwater in sheltering the coasts from heavy rollers. These kelps now have a considerable commercial value as a source of alginin, much used in the textile and confectionery industries.

The rate of production of some seaweed beds has been measured. Net production is the resultant of growth, less attrition due to the buffeting of the waves, to the action of animals, to respiration, and to loss of material in the form of eggs and sperms which are produced in immense numbers.

As among land plants, the rate of photosynthesis increases logarithmically with increase of temperature, over the range 0 to 20°C. Above 30°C, the rate of production may fall as the enzymes are injured. Respiration generally doubles for each 10°C rise in temperature, but is only about one-twentieth of the rate of production by photosynthesis. However, the short period of the year available for photosynthesis in the Arctic, where seaweeds grow in profusion, requires the storage of much food reserves to provide for respiration through the long polar night.

In favourable conditions seaweeds can double their weight in a few days, and can equal the rate of carbon fixation of the most productive plant communities that have been measured, at about 20 gm.C/m^2/ day.

But the rate of production diminishes as the weed grows denser and shades out its own fronds, until the rate of photosynthesis is little greater than the loss by respiration. Clearly, an artificial thinning out of the growing weed would very greatly increase the rate of production, and this is the basis of seaweed cultivation. Nevertheless, a rocky beach near Plymouth, England, carried a seaweed cover varying from 5.7 kg./ m^2 near high water level to 14.3 nearer low water level.[12]

Seaweed Cultivation

Seaweed cultivation was once widespread in the Pacific area, but is now almost confined to Japan and China. The brown seaweeds *Undaria* and the familiar oarweed *Laminaria* are the species chiefly cultivated and about 60,000 tons of the former and 100,000 tons of the latter, dry weight, are produced annually. Cultivated *Laminaria* can produce 2.5 metric tons dry weight per hectare, or 250 gm./m^2, and cultivated *Undaria* about 2 kg./m^2.

Originally, the culture of these brown seaweeds did not go beyond strewing the sea bed with stones for the attachment of the naturally-produced young plants. But nowadays artificial collecting substrates such as strings and nets are placed for the attachment of the sporophytes. When well covered with small plants, the strings are attached to bamboo rafts, and bags of fertiliser, especially nitrogen fertiliser, may be hung among them to stimulate growth. Fronds of cultured *Laminaria* can grow to over three metres in four months, and can then be harvested and dried for the market.[13]

More extensive is the culture of the red alga *Porphyra* in Japan. This red seaweed is common in Europe and North America, but is little used as food there. It grows on rocky shores often on larger brown seaweeds. But in Japan the crop of cultivated *Porphyra*, which is sold as *aonori* in thin dry sheets, is about 120,000 metric tons. Demand exceeds supply, and the product has a very good analysis, including thirty to fifty per cent of nutritious and digestible protein as well as mineral and trace elements. The thalli are grown on long sheets of netting supported horizontally in shallow water, and are plucked at intervals throughout the growing season; the fronds regenerate for repeated plucking.

But these valuable seaweed cultures need much manpower, and are also threatened by the massive pollution of the Japanese coastal waters, especially with the dangerous heavy metal residues which have already poisoned large numbers of people there.

Pelleted seaweed meal is used as a stock feed supplement, because of its valuable content of trace elements, and there is also now a considerable industry making alginic acid from brown seaweeds. This material is used in the textile and confectionery industries. But even this usage is trifling compared with the great production of seaweeds on the rocky shores of the world. Their indirect value is very great, as a source of organic detritus, which plays such an important part in the feeding of so many marine animals.

The Infauna of Seaweed Beds

Lichens growing on rocky beaches are a combination of an alga and a fungus; they are successful plants and offer shelter under their flat thalli. In Devon, for example, a lichen growing on an intertidal rock was found to be harbouring in its fronds and holdfast a fauna of small animals equivalent to 274,000 per square metre of which 189,000 were bivalve molluscs and 57,000 isopods.[12]

On many tropical beaches, the brown seaweed *Sargassum* may grow sparsely; but the fronds, festooned with small grape-like bladders break away and drift about the oceans, still growing and sheltering a large specially adapted fauna showing a striking mimicry of the fronds.

Seaweed beds have a dense infauna which finds food and protection among the fronds and holdfasts. The brown seaweed *Pelvetia*, which can cover the rock with 5.5 kg./m^2, sheltered an average of 2,409 organisms per square metre, mainly amphipods and isopods. Lower down the shore, *Fucus spiralis* sheltered 5,630 organisms per square metre, and lower still, the bladder wrack, *Fucus vesiculosis*, the joy of children with its snapping and popping bladders, covered rock to the extent of 10 kg./m^2 and had in its fronds nearly 46,000 copepods, acarines, and small gastropod snails per square metre.[12]

The densest growth of all was that of the *Ascophyllum* and *Polysiphonia* beds, at 14.3 kg. weed/m^2, in which lived the fantastic fauna of 202,700 animals per square metre, chiefly ostracods, copepods, acarines, small snails and nematode worms. Below the level of ordinary spring tides are the vast beds of oarweeds, *Laminaria*, whose holdfasts alone sheltered 72,000 polychaete worms, nematodes, copepods, cirripedes and amphipods per square metre. A more recent estimate of the animal population of *Laminaria* holdfasts is 1,250 animals per holdfast, of 387 species from most of the invertebrate phyla.

Rock-Dwelling Filter Feeders

Clearly, the seaweeds of rocky shores are not only very productive, but also give food and shelter to hundreds of thousands of little animals living out their lives in the kelp jungles. They feed on each other and on the by- and breakdown-products of the seaweeds. In turn, these creatures will be food for many carnivorous animals and filter feeders.

These filter feeders include the barnacles, efficient little creatures which sweep the water with their hairy abdominal appendages for edible fragments. Other filter feeders include the sponges and many kinds of worms, which live in limy tubes and thence spread out widely their circle of brightly coloured tentacles garnished with cilia which entangle small particles, especially of plankton; and the sea squirts, which live sedentary lives attached to the rocks or to other organisms, passing a strong stream of water over their 'gills' which also act as efficient filters. Animals which get at least a part of their food by filtration include the cushion starfish, *Porania pulvillus*, whose cilia on the underside fan food particles towards the central mouth. Some sea cucumbers entangle suspended particles, alive and dead, on their sticky tentacles, and the crab *Porcellana* sweeps the water with its hairy appendages.

Oysters and Mussels

The most important filter feeders are the lamellibranch molluscs, especially the mussels and oysters, which grow attached to rocks and stones and mangrove roots, and can close their shells at low tides. A large mussel can filter the suspended matter from ten gallons of seawater daily, sorting out the inedible and rejecting it, and swallowing the rest. It is a most efficient animal, the more so since it has recently been shown that the gills and mantle cavity of mussels can absorb and concentrate dissolved organic matter such as glucose and amino-acids by many hundreds of times, so that a significant part of its diet can come from organic matter in solution. Further, very small particles or organic matter can be caught and digested on the gills, which secrete important amyloclastic digestive enzymes, and finally passed on into the 'hepatic caeca' where digestion and absorption are completed.[14]

Mussels occur in clusters, often covering very large areas, on rocks, piles, buoys and on beds of stones and gravel, which are exposed at low tides. A mussel bed is very productive, producing edible meat at 1,000 kg./ha., a higher rate than almost any system. The young are produced in vast numbers, and spend many days and weeks drifting with the tides and currents as they develop; they are then a very acceptable food item for many planktonic and filter feeding shore animals. To give the young a chance to colonise new habitats, and to reduce competition with the adults, the first settling stage, the so-called plantigrades,[15] settle temporarily on filamentous seaweeds and drift about on these; they may let go and cling to the underside of the surface film with their byssus threads. But ultimately they find a suitable substrate and besome sedentary animals, attached to the substrate by their horny and strong byssus threads.

Enemies

They have enemies, notably the shore crabs and the dog whelk; the whelk drills holes in the shell and scrapes out the flesh with its long toothed radula. One of the worst enemies is the starfish, which crawls onto its victim and enwraps it with the stomach, which can be turned inside out. Weakened by suffocation, the mussel cannot resist the pull of the sucker covered legs of the starfish, and when the shell has been opened, the starfish digests the flesh externally, finally absorbing in its stomach the produce of the digestion. Starfish are carefully weeded out from oyster beds by hand or by dragging tangles of rope ends over the beds. The true whelk, *Buccinum*, attacks shellfish in a different way. It forces the edge of its own shell into the open shell of its prey, which it can then kill and digest at its leisure.

Oysters were cheap at one time (Dickens' plebeian characters could afford them) but are now of course valuable and expensive. Mussels are also increasing in popularity and price. For more than a century oysters have been cultivated, and improved crops obtained by spacing out the growing shellfish. Specially prepared substrates, such as cultch, or old oyster shells, are strewn where the oyster 'spat,' or metamorphosing larval stages, will settle on them. These spat collectors are then relaid in deeper water, and well spaced out, so that each oyster can have a better access to food. Next, artificial spat collectors were introduced; the spat were collected on curved tiles coated with a mixture of sand and lime, and the tiles were built up into nests raised well above the ground. When they are seen to have collected a good load of settled spat, the lime coating is scraped off, thus removing the young oysters without harming them; they are then relaid in boxes well spaced out for fast growth.

More recently, both oysters and mussels are being cultivated by the 'hanging drop' method. The spat collectors in the case of oysters, and ropes in the case of mussels, are hung from rafts or frames, where they becomes well seeded with young shellfish. Then the rafts are removed to deeper water and the oysters thinned out by a wider spacing of the spat collectors on the ropes. By this arrangement, the shellfish are able to filter planktonic food away from the whole water column, thus allowing a much denser population; and by their suspension above the sea bottom they are safe from such pests as starfish and whelks and whelk tingles which cannot swim. The rate of production of mussels in the Spanish culture industry, by the hanging drop method, is as high as 300,000 kg. of meats per hectare, a rate of production probably unmatched.

Browsers and Scavengers

Browsers are important on rocky shores; among these are the periwinkles, limpets, chitons, which crawl over the rocks scraping off with their file-like radulas and swallowing the algal felt and its infauna. They play an important part in helping to clear oil spills from rocks. The oil is not digested, but is passed out as faecal pellets, which, being finely divided, are more easily oxidised and disposed of.[16] It has long been known that epiphytic plants and animals will grow on an oil film, and these browsers scrape off the oil film for the sake of these epiphytes. The sea urchins are also scrapers of weeds from the rocks, but in their case they use their pointed teeth for the puspose.

The scavengers include the crabs, lobsters and prawns, all valuable animals. The common shore carb is food for octopi and some fishes, while the crab, lobster and crawfish are caught in traps, but usually below low tide mark.

Predators

The rich fauna of tidal shores supports many predators, and also attracts many others, normally resident below tide marks, to follow the rising tide to feed on the mud and sand flats, and among the kelp beds.

The many kinds of sea anemones, both simple and colonial, are among the most important resident predators. Their ring of tentacles are well adapted to stun and then catch small animals with their paralysing poison cells. Many worms such as the sea mouse, *Aphrodite*, feed on other worms; and some species of the rag worm, *Nereis*, have protrusible chitinous jaws which can be used to snatch small prey. The ribbon worms, or nemertines, can swallow quite large invertebrates, having a boa constrictor's ability to distend the stomach without inconvenience.

Many molluscs feed on other molluscs, and so do many starfish, as has been mentioned.

On the rising tide come the fishes and cuttlefish, to seize the appendages of unwary worms and clams, and take their fill of shellfish. The grey mullet include the meiofauna in their diet; and predatory fish include the conger eel, wrasses, plaice, sole, bass and cod.

Young plaice, for example, migrate up and down the beach with each tide, though feeding is mainly done during daylight hours irrespective of the tide.[17]

The cephalopods are active predators in the rocky environment, as they are elsewhere. But it is interesting to note that the octopus will recognise as prey a crab running naturally over the sand or backed in fighting posture against a rock; but will not recognise a crab in an unnatural situation, as when offered suspended by a thread. In fact, the octopus may try to blow it away with a blast from its jet. But as soon as the crab is lowered to the sand and begins to run, it is recognised immediately and seized.

The intertidal zone is less important than it used to be, in the more advanced countries, because of the rise of offshore fisheries. But beach seining is still a dominant source of fish in many tropical countries: the huge seines, half a mile long, used in countries such as Ceylon, Malaysia and Arabia, are among the chief sources of fish there. Such methods are largely abandoned in Britain, except for high-priced fish such as salmon: large crews are necessary and labour hard to find. Shellfish collection is on a small scale in spite of the abundance of good shellfish; in other lands, highly priced shellfish such as the abalone and many kinds of clams are still an important industry. In Italy and the West Indies, 'sea-eggs,' the ripe roes of sea urchins, are a well-priced luxury.

Mussels and rock- and mangrove-oysters and octopus are still hand gathered; but the limpets (*Patella* spp) are too tough for modern tastes. But as a food of last resort, they were the mainstay of the ageing inhabitants of the island of Samson, in the Isles of Scilly, after the tragic death of the younger men in a sea accident.

The valuable food fish which follow the rising tide are fished for by handline and 'spiller' or staked longline. Small inshore trawlers will come with the tide to fish for shrimps and prawns, and for soles, plaice and cod. The bass is a fine sporting fish in the surf.

In past years, a variety of fixed fishing gears took fish which ascended the beaches on the rising tide. Davis gives particulars of traps of wattle, stones and netting set for sprats, herrings and flatfish.[18] Salmon stake nets are well known, but similar nets are set to catch any other fish available. In Passamaquoddy Bay in eastern North America and in some English bays, stake nets ('herring hangs') made large catches of herrings and sprats, which are removed at low tide.

NOTES

1. Russell, F.S. and Yonge, C.M. 1927. *The Seas.* London, Frederick Warner & Co. Ltd., 1927.
2. Davidson, P.E. 1967. 'A Study of the Oystercatcher (*Haematopus ostralegus* L.) in relation to the fishery for cockles (*Cardium edule*) in the Burry Inlet, S. Wales.' *Fish. Invest. Ser. II,* xxv (7), London, HMSO, 1967.
3. Korringa, P. 1967-8. 'Estuarine Fisheries in Europe as affected by man's multiple activities.' In *Estuaries* vol. 2, Ed. G. Lauff. Amer. Assn. Adv. Sci. Washington D.C.: 658-63.
4. Davis, F.M. 1925. Quantitative studies on the fauna of the sea bottom, No. 2. 'Results of the investigations in the southern North Sea.' *Fish. Invest. VIII* (4) London, HMSO, 1925.
5. MacIntyre, A.D. 1970. 'Range of biomass in intertidal sand, with special reference to *Tellina tenuis.*' *J. Mar. Biol. Assn. U.K.* (1970): 561-75.
6. Harris, Roger P. 1972. 'The distribution and ecology of the interstitial meiofauna of a sandy beach.' *J. Mar. Biol. Assn. U.K.* (1972) 52: 1-18 and 375-9.
7. MacIntyre, A.D. 1969. 'Ecology of marine meiobenthos.' *Biol. Rev.* 44: 245-90.
8. Petersen, C. Johs. 1918. 'The sea bottom and its production of fish food.' *Danish Biol. Sta. Repts.* 25 (7), 1918.
9. Qasim, S.Z. and Bhattathiri, P.M.A. 1971. 'Primary production of a sea grass bed on Karavatti Atoll.' *Hydrobiologica* 38: (1) 29-38.
10. Fenchel, T. 1970. 'Studies on the decomposition of organic detritus derived from turtle grass.' *Limn. Oceanogr.* 15: 14-20.
11. Kanwisher, John W. 1966. 'Photosynthesis and Respiration in some seaweeds.' *Some contemporary studies in Marine Science* (1966): 407-40.
12. Colman, J.S. 1940. On the faunas inhabiting intertidal seaweeds. *J. Mar. Biol. Assn. U.K.* 24: 129-83.
13. Bardach, John, Ryther, John and McLarney Wm. O. 1972. *Aquaculture.* New York, Wiley Interscience, 1972: 790-814.
14. Pequinat, E. 1973. 'Assimilation of amino acids and glucose by *Mytilus.*' *Mar. Biol.* 19 (3): 227-44.
15. Bayne, B.L. 1964. 'Primary and secondary settlement in *Mytilus edulius.*' *J. Anim. Ecol.* 33: 513-23.

16. Smith, J.E. ed. 1968. *'Torrey Canyon' pollution and marine life*. Cambridge University Press.
17. Gibson, R.N. 1973. 'The intertidal movements and distribution of young fish on a sandy beach with especial reference to the Plaice (*Pleuronectes platessa* L).' *J. Exp. Mar. Biol. Ecol.* 1973, 12: 79-102.
18. Davis, F.M. 1936. *An account of the Fishing Gear of England and Wales*. 3rd edn. Fish. Invest. London. XV. No. 2. 1936.

Chapter 12 PRIMARY PRODUCTION IN THE OCEAN

In Chapter 11, the very high rate of primary production by the seaweeds in the intertidal zone was estimated at twenty times that of the phytoplankton in the ocean. But this high production is confined to less than one per cent of the ocean's surface; whereas the lesser rate of phytoplankton primary production occurs over an area equal to two-thirds of the surface of the entire globe. The primary production of the freshwaters, however important locally, is negligible on the global scale.

The Organisms Responsible for Primary Production in the Oceans

In temperate seas, the diatoms are the principal producers, at least in the spring, and sometimes great blooms are formed. I recall a very still and sunny day far out at sea off the southwest of Ireland in March, when the proliferation of the diatoms was so rapid that a scum of diatoms formed an oily layer on the ocean surface, and a sample could be scooped up in a bucket and identified. In the Arctic and Antarctic also, diatoms are the chief producers. But in the warmer months in temperate climates, and usually in the warm waters of the subtropics and tropics, the diatoms are replaced by flagellates, and especially the peredinians.

In the Mediterranean, for example, the chief producers are the Coccolithophores, a family of small flagellates, and the diatoms play a negligible part.[1] Though these coccolithophores are naturally most abundant in the upper, photic, zone, at depths between zero and fifty metres, they were also very abundant at 250 to 800 metres far below the photic zone, and were still abundant at 1,000 to 2,500 metres. As these deep living cells were healthy, they must have been living at the expense of particulate matter, or on dissolved organic matter, and in either case saprophytically.

It has been known since the turn of the century that very minute flagellates, first studied as caught on the very fine filtering apparatus of salps, are an extremely important part of the phytoplankton.[2] They may in fact produce many times more organic matter than all the larger plants, including the diatoms.[3] These flagellates are small enough to pass through the meshes of the finest plankton nets, and are collectively called the 'nanoplankton.' They have already been encountered in the phytoplankton of rivers (p. 27). They are important not only as primary producers, but as food for many filter feeding organisms.

Owing to their small size and delicate structure, they are hard to study, and are not so well known as their importance deserves. They can live as plants or as animals indifferently. Though they may be only a few thousandths of a millimetre in length, they can show some intelligence and power of discrimination. One such flagellate, seven thousandths of a millimetre long, possesses a yellow pigment and could live like a plant, performing photosynthesis in sunlight. But it could also wrap itself round a bacterium and digest it. Offered a mixture of bacteria and very finely divided graphite, it had no problem in accepting the bacteria and rejecting the inedible graphite.[4]

A very interesting comparison has recently been made between the plant constituents of samples taken from the surface mixed layer of water, respectively off the east coast of south America, the southwest coast of Africa, in the Gulf of Mexico, and in the rich upwelled water off Peru.[5] In these four widely distributed samples, diatoms were the dominant constituents of the phytoplankton only in the very rich upwelled water off Peru. In the other three situations gymnodinians (peredinians) and other microflagellates were dominant in the phytoplankton.

The Rate of Primary Production

Measurements of the rate of primary production in the oceans are obviously of great importance, so early attempts were made in 1916-7 by Gaarder and Gran in the Oslo Fjord.[6] These workers filled glass bottles with the seawater they were investigating and inoculated them with a known number of phytoplankton organisms. The oxygen content of the water was also determined. Then some of the glass bottles were blacked out with opaque paint, and all were lowered to the required depth in the sea. After an interval of some hours, the clear bottles in the sunlight will have increased their content of dissolved oxygen by the photosynthesis of some of the dissolved carbon dioxide. But the blacked out bottles would have been unable to carry out photosynthesis, while the contained plant cells will have been actively using up oxygen in respiration. There will have been a net loss of oxygen in these blacked out bottles, a loss which shows how much oxygen is being used up in respiration, a quantity concealed in the lighted bottles because the carbon dioxide respired will immediately have been reduced to carbohydrate with the evolution of oxygen. To get total oxygen production, this quantity will have to be added to that found in the lighted bottles. The method has been applied also to fresh waters in rivers and lakes.

The method was developed further by Marshall and Orr, in the Firth of Clyde.[7] It has been for many years a standard method of estimating primary production. For if the amount of oxygen produced in a given time is known, the amount of carbon reduced from the dissolved carbon dioxide can be calculated, and thence the rate of primary production. But the success of this method depends on Winkler's method of oxygen estimation; this method is not sensitive enough to measure the minute quantities of oxygen produced in nutrient-poor oceanic waters.

Though now replaced by a better method about to be described, very useful information was got from the light and dark bottles. For example, it was shown that there is a range of primary production in seawater as wide as from 5 to 1,000 gm./m^2/annum.

Meanwhile, W.R.G. Atkins, at Plymouth made an attempt by a different method.[8] Though seawater is well buffered, so that local changes in pH are quickly corrected, it does undergo measurable changes as the carbon dioxide in the water is reduced, making the water slightly less acid. By following the changes in pH by the then novel methods available, he calculated that about 1.4 kg./m^2 of algal material was produced in the western English Channel during the months of growth from December to July. This wet weight of algae contained about fifteen per cent of carbon; whence the rate of production is about 210 gm.C/m^2/annum. By a different method, namely, the rate of utilisation of the dissolved phosphate, Atkins arrived at the same figure.[9] Recent estimations by the Carbon ^{14} method (described below) have given values for the western English Channel of about 140 in 1964, rising to 212 in 1970. But in 1922, the western English Channel was in a stage of high productivity to which it is now returning after a period of years of low productivity (p. 161), so Atkin's pioneer estimations agree well with recent results by newer methods.

The Carbon ^{14} Method

Just before the sailing of the Danish research vessel 'Galathaea' on her great expedition round the world in 1946, the Danish biologist Steeman-Nielsen was able to get one-quarter gram of barium carbonate, in which a very small part (about 0.5 per cent) was in the form of the radioactive isotope C^{14} in place of the usual C^{12}. He had time to devise a technique to use it.

Plants, in the process of photosynthesis, do not distinguish between the two isotopes of carbon, and so, if a known quantity of radioactive $Carbon^{14}$ is added to a known quantity of seawater whose rate of primary production is to be measured, the plants will use the radioactive isotope in the same way as the ordinary carbon; so that a determination of the amount of radioactive carbon fixed in photosynthesis will represent the whole of the carbon. This amount is simply determined by the use of a Geiger counter; all that is necessary is to filter off the sample, after the given period of photosynthesis, with a collodion membrane having pores of 1/5000 mm. fine enough to hold all the cells concerned with photosynthesis, and then determine the radioactivity of the precipitate. Though it was later found that a correction has to be made because of selfabsorption and scattering in the active carbon, the method is now in worldwide use, and can give reliable results in a short time.[10] The shorthand method now used is described by Steeman-Nielsen in the popular account of the 'Galathaea' voyage.

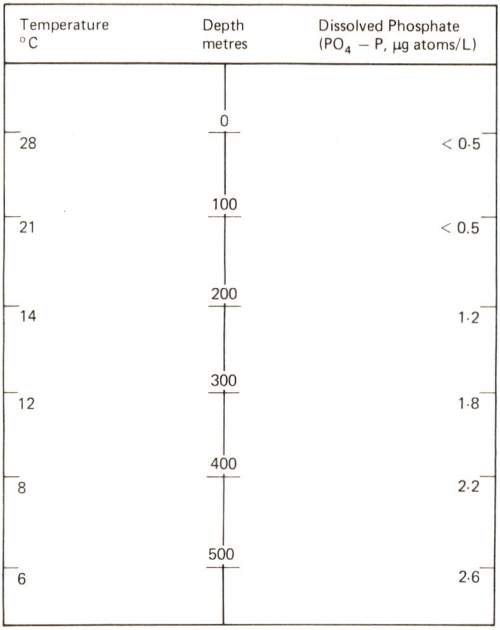

Figure 6 Stratification of oceanic water; a warm layer of water, denuded of nutrient salts, overlies deeper colder water in which nutrient salts become more abundant.

The Global Annual Primary Production in the Oceans

The global production of the oceans was estimated by Steeman-Nielsen and Jensen at about 16,000 million tons (16×10^9) of carbon annually, which is equivalent to about 33,000 million tons of carbohydrate.[11] This is about the same as the estimated primary production of the land areas of the globe. But since there is three times as much ocean area as land area, it follows that the land is about three times as productive as the oceans, area for area. A value of about 20,000 million tons of carbon is now generally accepted (p. 145).

The excretion of Organic Matter

In fact, the above rate of oceanic production is likely to have been underestimated; for it is known that the phytoplankton may excrete into the surrounding water a considerable proportion of the carbon they assimilate, and that this proportion may increase from the shore seawards. Thus, it was found to be less than seven per cent in the estuaries of Georgia, in the southeastern United States, and less than thirteen per cent in the coastal waters. But in the western Sargasso Sea it was found to be of the order of forty-four per cent.[12] The carbon material excreted will plainly not be included in the results of the $Carbon^{14}$ technique, and introduces an error in estimates of oceanic primary production. But this excreted organic material plays an important part in the cycle of marine life.

Regional and Seasonal Variations in Primary Production

Since the amount of sunlight is about the same over large areas of ocean, and since carbon dioxide is never likely to be in limiting supply, variations in oceanic production, which are great, are chiefly due to variations in the availability of nutrient salts, chiefly phosphate and nitrate. These are present in virtually unlimited quantity below the photic zone of about 250 metres; but the formation of a stable warm layer of water at the surface, overlying thousands of metres of colder, heavier water and separated from it by a thermocline (see p. 39), prevents this rich source of nutrients from reaching the photic layer where it could contribute to global primary production. Only where water movements break down the thermocline does this rich water upwell, with very notable effect.

This is so well known and so important that, for decades, the minds of men have turned to the possibility of bringing this deep water to the photic layer artificially. But the stability of the ocean stratification is very great, and enormous quantities of energy, beyond our present imagining, would be needed to secure a worthwhile overturn. Moreover, the artificially upwelled water would still be colder and heavier than the surface water, and would tend to drift downwards again. Only in confined shallow areas close to deep water would such an artificial overturn be possible. It has been suggested that nutrient-rich deep water, discharged into confined areas, could increase primary production by a large factor. Such an atoll lagoon as Kwajalein, for example, could produce, by deliberate culture methods in the enriched water, enough marine animal protein for ten million people annually.[13] Experiments at St Croix, in the American West Indies, have shown that seawater drawn from a depth of 700 metres can stimulate a rate of primary production of 2 mg.C/m^2/day, which is about the same as the rate of production of the Peruvian Current.[14]

Upwelling

A natural upwelling of the deep oceanic water can happen in a number of ways. For example, it occurs when prevailing offshore winds blow the spent surface water away to sea and drag up deeper water to replace it. The waters off southwest Africa, where the northgoing Benguela Current diverges slightly from the coast, and off Peru, where the waters of the Peruvian or Humboldt Current do the same, and off northwest Africa, are all notable cases. Though the water upwells from only a depth of a few hundred metres, so that the great reserves of the ocean are hardly touched, a vast proliferation of the phytoplankton gives a high rate of primary production. The 'Galathaea' found values of 3.8 gm.C/m^2/day off southwest Africa; in such nutrient-rich water the rate of primary production is usually between 2.0 and 3.0 gm.C/m^2/day. Hobson *et al* found rates of 9 gm.C/m^2/day in the Peruvian upwelled water; but only 0.1 gm.C in the spent water of the Gulf of Mexico.[5]

In the area of upwelling off the coast of northwest Africa, primary production was running at 1.12 to 3.35 gm.C/m^2/day.[15] As the north-east trade winds drive this enriched water westwards across the Atlantic, its riches are gradually used up in supporting successive generations of phytoplankton which live out their short lives and then die, their remains sinking in most cases below the photic layer. By the time the water has drifted to the west Atlantic near the island of Barbados, its rate of primary production has fallen to 0.29 gm.C/m^2/day, and an estimated annual net rate of production of 105 gm.C/m^2.[16] In the Sargasso Sea, in the midst of the northern Atlantic gyre, the rate of primary production may be only 0.16 gm.C/m^2/day. Steeman-Nielsen's own values for the Sargasso Sea were even lower, at 0.05 gm.C/m^2/day.

Turbulence

Where ocean currents pass each other turbulence is set up and, especially where two currents diverge, nutrient-rich deeper water is dragged to the surface. A belt of oceanic water of high productivity, for example, is found in the Pacific, along the zones of the Equatorial currents and countercurrents. Steeman-Nielsen found values of 0.6 gm.C/m^2/day here, as compared with values of about 0.15 in the mid-Pacific generally. Turbulence can also be set up when ocean currents flow over irregularities on the sea floor, such as banks and canyons, and when they are deflected up the rising slopes of the continental shelves. The productivity of fishing banks is famous, and upwelled deep water contributes to the fertility of the continental shelves.

The Southern Ocean

An important case of upwelling happens all round the Antarctic Continent. There, the sea freezes, and the ice floes and bergs are composed of fresh water, leaving the sea saltier as well as cold. This heavy water sinks, and is replaced by deep but lighter water rich in nutrient salts. Consequently, the southern ocean which surrounds the Antarctic is the most fertile of all the oceans. It produces vast crops of diatoms, which nourish a zooplankton rich enough to support the 800,000 baleen whales estimated to have lived and migrated there before the days of mechanised whaling, besides great populations of seals and penguins. But more than this: even the full range of daylight in the southern summer is insufficient to use up all the plant nutrients available, so that unused nutrients are carried north in currents which join the Benguela and Humboldt Currents, and in upwelling these are a source of wealth and prosperity to South Africa and Peru.

The Continental Shelves and Coastal Waters

These waters receive the products of land drainage as well as contributions from upwellings at their seaward slopes; and the shallow depth, of 200 metres and less, guarantees that dead material will not fall far below the photic zone, and will soon return to circulation. So high values for primary production are found in coastal waters and over the continental shelves; for example, 0.71 gm.C/m^2/day in the Skagerrak, 0.20 in the central North Sea, 0.72 in the southern North Sea, which receives the treated sewage of London and other cities, and 0.47 in the western English Channel. In the Straits of Malacca, a value of 0.53 was found; and 0.78 and 0.49 in the Gulf of Thailand. Such high rates of primary production are usually associated with rich fishery resources.

Production in Temperate Climates

Whereas in the tropics primary production is not influenced by seasons, but is controlled by the availability of nutrient salts, in temperate climates, there is a regular seasonal cycle of exhaustion and regeneration of plant nutrients. This is essentially the same as was discussed in Chapter 4 in lakes. In winter, there is a complete mixing of the water by the winds, but in spring the increasing sunlight triggers off the annual growth of the phytoplankton. Even to the naked eye, the surface of the ocean can show a change of colour due to the dense populations of diatoms. This rapid multiplication of the phytoplankton continues

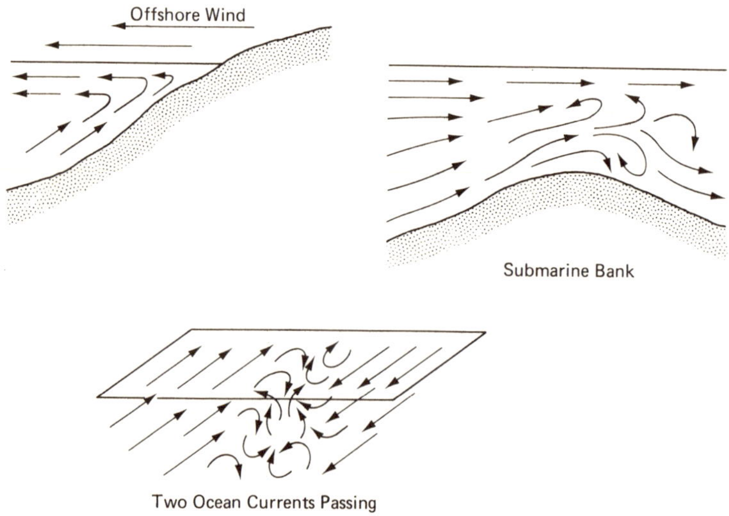

Figure 7 Some ways in which oceanic waters are mixed

until the supply of nutrients in the surface layer is exhausted. Meanwhile, the growing power of the sun has warmed the surface layer of the sea, which now overlies the deeper colder and heavier water. A sharp discontinuity can develop, called the thermocline, which can hinder mixing of the two layers. Thus, as summer draws on, the surface layer, above the thermocline, may be spent of nutrients, while there is still plenty below the thermocline. The phytoplankton then tends to die down, until the falling temperatures of autumn break down the thermocline and mix the deeper water, bringing with it renewed nutrient salts. A second efflorescence of the phytoplankton follows; but it is necessarily less intense than the spring one, for not only is the temperature falling, but the daylight is waning on which photosynthesis depends.

Meanwhile in the spring the zooplankton also develops, feeding and growing on the rich supply of the phytoplankton. This zooplankton includes not only permanent members, but temporary members such as the young stages of many invertebrates and fish. This increasing population of zooplankton grazes down the phytoplankton, and is recognised as a limiting factor in the development of the plankton. A secondary outgrowth of the zooplankton also follows the secondary phytoplankton growth, and, like it, is short.

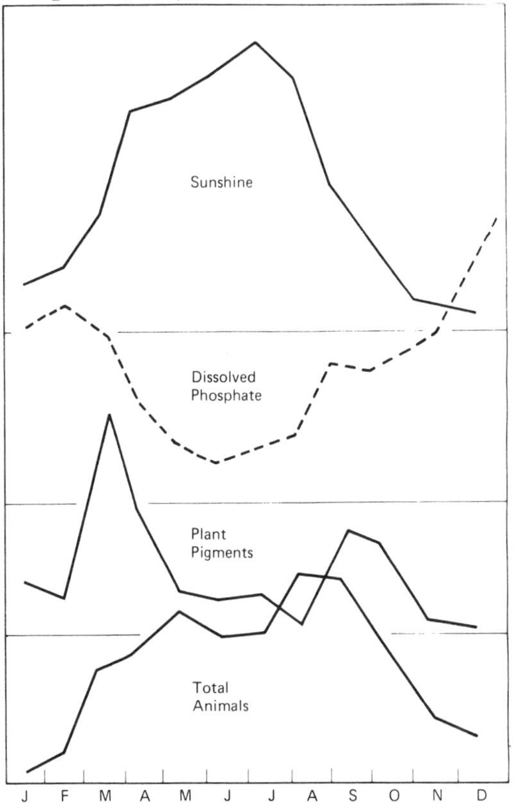

Figure 8 The relation between sunlight, dissolved nutrient salts, the growth of the phytoplankton, and that of the zooplankton, in the western English Channel.

In the Arctic, the continuous daylight allows of an enormous and complete use of the nutrient salts which have become available, and this single outburst of primary production has to tide the fauna and flora of the Arctic through the long dark months of winter, when there is no light for photosynthesis even though nutrient salts may be in plentiful supply. The Arctic brown seaweeds have to cope with the same problem (Chapter 7).

Excretion of Organic Matter by the Phytoplankton

It has already been said that as much as fifty per cent of the new organic material synthesised by green plants in the ocean may be excreted into the seawater. The study of Hobson *et al* gives more information about this. They found that the concentrations of dissolved organic carbon at the four widely-spaced stations they examined varied from 0.5 to 1.5 g./litre in three areas, but was 4.5 in the very fertile ocean off Peru. No doubt this was derived from the very rich diatom flora of those waters. But carbon also occurs in the nonliving particulate form. Such particles can be derived directly from dissolved organic matter. Hobson *et al* found that the concentrations of particulate organic carbon were higher in nutrient-rich than in nutrient-poor water, pointing clearly to its origin in the phytoplankton.

Organic particles will form in undisturbed membrane-filtered seawater under a variety of conditions; if these particles are filtered off, another lot will form.[17] In the Strait of Georgia, British Columbia, for example, particulate organic matter may be a significant three to thirty per cent of the total particulate matter present in the ocean. The carbon present in the bodies of the phytoplankton organisms may be only twenty to fifty-five per cent of the total except in areas of large standing crops of phytoplankton, where it could rise to seventy to ninety per cent.

These particles, which often take on a plate-like appearance, form a good substrate for the growth of bacteria, and thus begin a saprophytic food chain which parallels the more important food chain which starts with the phytoplankton itself. Enriched organic particles can be food for the zooplankton,[18] and as bacterioplankton are a major source of food for filter feeders.[19] In certain conditions, the production of bacterioplankton may greatly exceed autotrophic production by the phytoplankton, so that soluble organic matter transported from temperate climates may have to be drawn on for the extra energy required.

The number and biomass of marine bacteria changes by several hundred times between those present in the sediments of lagoons and those present on the floor of the deep ocean, corresponding to changes in the biomass of the bottom fauna (p. 187).

The part played by dissolved and other nonliving organic matter in the oceans is very important and is under close study. It has been estimated that the oceans may contain an average of 0.35 to 0.7 mg./litre, which would add up to the staggering total of 5 to 10×10^{12} metric tons in the whole oceans.

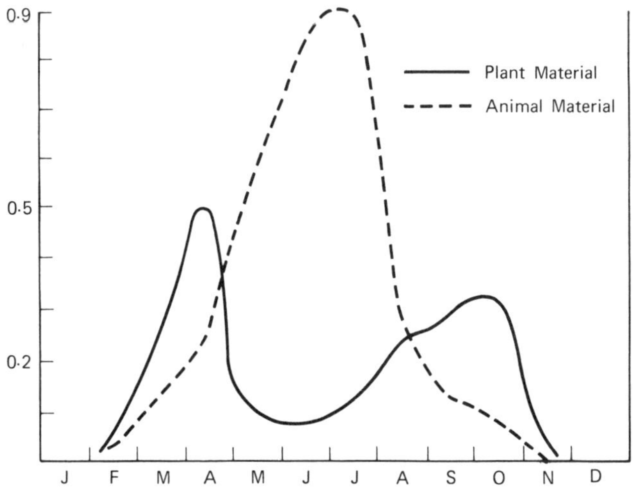

Figure 9 The succession of plant and animal plankton shown in Figure 8 is reflected in the stomach contents of the Pilchard, *Sardina pilchardus*, a plankton feeder, in the English Channel.

The Oceanic Food Chain

The primary production of the oceans, estimated at something like 20,000 million metric tons of carbon annually, can obviously take place only in the photic layer of the ocean, say the topmost 100 metres of oceans thousands of metres deep. From there, it becomes distributed throughout the ocean by way of animal life; it is the support of the 'creeping things innumerable, both small and great beasts' of the great and wide sea.

One schematic distribution of this largesse is as follows[20]:–

Production of Phytoplankton	20,000 million metric tons/C.
Whence production of zooplankton	2,000 million metric tons/C.
and of zooplankton eaters	200 million metric tons/C.
and of predators	20 million metric tons/C.

In the steps shown in the chain, there is a loss of ninety per cent of the matter (and energy) at each link. This is because the animals need about ninety per cent of the energy of the food they eat for their own maintenance, leaving only about ten per cent surplus which can be used for growth; and it is this which is handed down to the next step in the chain.

Since the newly-fixed carbon must be most plentiful in the photic zone, it follows that here will be the most plentiful supply of food for the grazing zooplankton, and that the further down into the depths from there, the less the food supply, and therefore the fauna, is likely to be.

Man makes but little use of this vast oceanic primary production. A few hundred thousand tonnes of seaweeds are used, and about fifty-five million tonnes of marine fish. If fish contain about ten per cent of carbon, we can take only 5.5 million tonnes of carbon from an annual harvest of 20,000 million tonnes, or about 0.027 per cent. This is a minute proportion compared with the use we make of the primary production of the land.

But whereas fruits, seeds, roots and herbage are easy and cheap to harvest, phytoplankton is patchily scattered in three dimensions and can be harvested, if at all, at great cost. There is a better prospect of using zooplankton, hypothetically only one-tenth as abundant as the phytoplankton; but results so far suggest that edible meal made from zooplankton is many times more expensive than meal of equal quality made from fish.

On the other hand, man takes about 1.2 per cent of the more easily harvestable zooplankton-eating fish, and a much higher percentage of the predators.

NOTES

1. Bernard, F. 1958. 'Données récentes sur la fertilité élémentaire en Méditerranée.' *Rapp. Cons. Exp. Mer.* 144: 103-8.
2. Lohmann, H. 1903. 'Neue untersuchungen über den Reichtum des Meeres an Plankton.' *Wiss. Meeresunters. Abt Kiel* 8: 1-86.
3. Murray, J. and Hjort, J. 1912. *The Depths of the Ocean.* London, Macmillan.
4. Swale, E.M.F. 1969. 'A Study of the Nanoplankton *Pedinella* by light and electron microscope.' *Brit. Phycol. J.* 4: 65-86.
5. Hobson, L.A., Mengel, D.W. and Barber, R.T. 1973. 'Primary productivity and pools of organic carbon in the mixed layer of the sea.' *Mar. Biol.* 19: 298-306.
6. Gaarder, T. and Gran, H.H. 1927. 'Investigations of the production of Plankton in the Oslo Fjord.' *Rapp. Cons. Exp. Mer.* 42: 1-48.
7. Marshall, S.M. and Orr, A.P. 1928. 'The photosynthesis of diatom cultures in the sea.' *J. Mar. Biol. Assn. U.K.* 15: 321-60.
8. Atkins, W.R.G. 1922. 'The hydrogen-ion concentration of sea water in its biological relations.' *J. Mar. Biol. Assn. U.K.* 12: 717-71.

9. Atkins, W.R.G. 1923. 'The phosphate content of fresh and salt waters in its relationship to the growth of algal plankton.' *J. Mar. Biol. Assn. U.K.* 13: 119-50.
10. Steeman-Nielsen, E. 1956. 'Measuring the productivity of the Sea.' *The Galathaea Deep-Sea Expedition*, 1950-2, London, Geo. Allen & Unwin.
11. Steeman-Nielsen, E. and Jensen, E.A. 1957. 'Primary Oceanic Production The autotrophic production of organic matter in the oceans.' *Galathaea Repts* 1: 49-136.
12. Thomas, J.P. 1971. 'Release of dissolved organic matter from natural populations of marine phytoplankton.' *Mar. Biol.* 11: 311-23.
13. Isaacs, J.D. and Schmitt, W.R. 1969. 'Stimulation of Marine productivity.' *J. Cons. int. Exp. Mer.* 18 (3): 287-95.
14. Bienfang, P. 1971. 'Taking the pollution out of Waste Heat.' *New Scientist*, 26 Aug. 1971.
15. Lloyd, I.J. 1971. 'Primary production off the coast of northwest Africa.' *J. du Conseil int. Explor. Mer.* 33 (3): 312-23.
16. Steven, D.M. 1971. 'Primary productivity of the tropical west Atlantic near Barbados.' *Mar. Biol.* 10: 261-4.
17. Sheldon, Evelyn and Parsons, T.R. 1967. 'Small particles in seawater.' *Limn. Ocean.* 12: 367-75.
18. Riley, Gordon, A. 1963. 'Organic aggregates in seawater and the dynamics of their formation and utilisation.' *Limn. Ocean.* 8: 372-81.
19. Sorokin, Y.I. 1971. 'Bacterial populations as components of oceanic systems.' *Mar. Biol.* 11: 101-5.
20. Gulland, J.A. 1968. 'Food Chain studies and some problems in World Fisheries.' In *Marine Food Chains*: 296-315. Edinburgh, Oliver & Boyd. 1970.

Chapter 13 THE PELAGIC ENVIRONMENT

The members of the phytoplankton must be able to maintain themselves near the surface, or at least in the photic layer, to be able to carry out photosynthesis. The rate of sinking of particles in the sea is in inverse proportion to the surface area. Thus, very small organisms such as bacteria and flagellates, and the smaller peredinians and diatoms, will sink so slowly that the rate is far less than the forced movements brought about by turbulence in the water. The larger phytoplankton organisms can secure the same result by growing long slender spines and wing-like extensions which greatly increase their surface area. It is notable that plankton organisms which live in warmer waters, which have a lower viscosity and would tend to accelerate the rate of sinking, are particularly liable to grow fantastic and beautiful appendages and processes. The larger phytoplankton can secure a like result by secreting oil droplets and other light materials. However, even with these devices, phytoplankton is often carried by turbulent water movements down below the photic zone, and this can sometimes explain observed anomalous rates of primary production.

There is evidence that the ocean's crop of phytoplankton varies from year to year locally; but it is unlikely that the ocean as a whole would be simultaneously affected.

Phytoplankton-Feeding Fish

Many species of fish, including some very important ones, feed directly on the phytoplankton, at least during a part of the year. They are members of the herring family, and include the menhaden of the eastern coast of America, the bonga, *Ethmalosa fimbriata* of tropical west Africa, and the oil sardine, *Sardinella longiceps* of the Indo-Pacific region. The first provides most of the fish meal and oil used in the USA, and the oil sardine is a dominant factor in the prosperity of the west coast of India. The anchovies are also in many cases predominantly direct feeders on phytoplankton. These include the *Engraulis capensis*, which shares and supplants the likewise mainly phytoplankton feeding *Sardinops ocellata* in the very important industrial fishery off southwest and south Africa. Most important of all, they include the *Engraulis ringens*, the anchoveta, of Peru and northern Chile, which is a phytoplankton feeder during most of the year, except in the months of April and May, when zooplankton is chiefly eaten.[1] As the table on page shows, fish which feed directly on the phytoplankton have an enormous amount of food at their disposal, and this is one of the factors which gives these clupeoids their very great productivity: they are only one step down from primary production.

The Anchoveta Fishery

In a normal year, the Anchoveta fishery is the greatest in the world, and provides about fifteen per cent of all the fish caught. It is used for the manufacture of fish meal, and its abundance can dominate the world price of this increasingly sought-after commodity. The Anchoveta not only provides millions of tons of fish; it also feeds millions of sea birds. It was estimated that in 1955 a bird population of about twenty-seven millions was eating about four million tons of Anchoveta.[2] For a century before the growth of the commercial fishery for Anchoveta, the excrement of these shags, collecting on the outlying desert islands where they roost and breed, was the main source of guano, a cheap and abundant fertiliser; while the more completely mineralised guano was at one time the world's chief source of nitrate fertiliser, the 'Chili saltpetre' of commerce.

But, at intervals of a few years, the region of upwelling of nutrient-rich and cold water of the Peruvian Current is invaded by warm water of tropical derivation, and then disaster comes. The anchoveta shoals remain deep in the cold water below the surface warm layer, inaccessible to both the diving birds and the encircling nets of the fishermen.[3] The birds scatter and emigrate, and starve. When 'el Nino' happened last, the population of these seabirds fell from sixteen million to about three to four million. In 1972, 'el Nino' struck again, and the catch of anchoveta fell from the usual eight to ten million tons to one million only. This failure played a large part in the unequalled rise in the price of fish meal, base of most animal feedstuffs, in 1972-3. This contributed to the rising price of meat, milk and eggs, causing social discontent the world over.

The Zooplankton

But the phytoplankton is mainly consumed by the zooplankton, a very large group of invertebrate animals which for the most part spend the whole of their lives in the pelagic zone of the sea, consuming the phytoplankton, and being consumed by larger members of the pelagic community such as the fishes and squids. The distinction between zooplankton and nekton, or organisms which can move actively, is arbitrary but convenient; there are many borderline cases, such as that of the Euphausiid crustaceans, which are active swimmers and yet include phytoplankton in their food and are usually included in the plankton. At all events, the very varied array of animals which comprise the zooplankton perform the essential function of distributing the primary production of the surface layers about the ocean, even to its deepest parts.

Though the food chain on page 145 indicates that the zooplankton will, globally, comprise only about ten per cent of the tonnage of the phytoplankton, this cannot always be seen locally. The relation between the plants and the grazers is complex. Like a well-grazed meadow, the grass may not seem to increase in quantity, though it will have produced many tons of fodder to cause growth in weight of the grazing cattle. The grazing of the zooplankton is the chief cause of the diminution of the phytoplankton. In an experiment, Wimpenny[4] found that two starved brine shrimps, a very small crustacean, in three days reduced a population of the diatom *Nitzschia closterium* from 3,830 cells/ml. to 2,006 cells. The phytoplankton itself is nourished by the excretions of the zooplankton, and by the products of the decay of the dead bodies of both plant and animal plankton. The zooplankton animals excrete phosphate, an essential plant nutrient, and the smaller the organism, the more rapid its metabolism.[5] Many years ago, Lohmann[6] showed that the nanoplankton, animals and plants so small as to pass through the finest meshes of nets, were as important as all the remaining plankton put together; and Hobson *et al*[7] showed that ciliates and very small crustacea such as copepod nauplii were the chief phytoplankton feeders. The microzooplankton in this case accounted for two to thirty per cent of the particulate carbon in the surface layer of the ocean, while the phytoplankton accounted for twenty to fifty-five per cent: the zooplankton were more than ten per cent of the phytoplankton.

Like the phytoplankton, the microzooplankton must have devices for floatation; many of them have very elongated limbs and appendages, especially in the tropics where the high temperatures of the water give it low viscosity. Again, as with the phytoplankton, there may be secretion of fats and of light ammonia compounds. But the larger zooplankters can swim vigorously, and so maintain their depth as required.

Temporary Members of the Zooplankton

Many invertebrates and fish have juvenile stages, usually very unlike the adult, which live among the phytoplankton and feed upon it while developing. They grow and metamorphose, often through many stages, and may then sink down into deep water, or sink to the bottom in shallower water, and thus carry with them material produced in the photic zone by the phytoplankton. This is a very important step in the distribution of primary production about the ocean.

All the plankton organisms have a short life, and at their death their remains fall as a rain into deeper water, decaying and releasing their nutrients as they fall. This process may last for weeks, and even months, contrary to the belief that the protoplasm decays rapidly.[8] But in this way the nutrient reserves of the waters are replenished, to be brought back into the photic zone for re-use by turbulent water movements or seasonal overturns.

Simple Zooplankters

The simplest members of the zooplankton are the ciliates, and the colourless flagellates which must feed on still smaller phytoplankters or on excreted organic matter. The foraminifers and radiolaria are minute drops of living matter resembling the familiar *Amoeba*, but protected by perforated shells of lime or silica. They swarm in myriads, and feed on small particles or organic matter, which they surround with protrusions of their bodies. Some harbour living plant cells unharmed inside their bodies, living in mutually beneficial partnership.

The jellyfish, both simple and compound, are motile and often support themselves with floats, below which trail their tentacles replete with poisonous stinging cells which stun and kill their prey, even small fishes. They often take serious toll of the zooplankton, and as they may occur by scores in every cubic metre of ocean, they can affect the future of the fisheries by their slaughter of baby fish. One specimen alone of the jellyfish *Cyanea* is capable of killing 15,000 baby fish.

The Crustacea

The crustacea are the dominant members of the zooplankton world. The copepods have been mentioned already; they possess food reserves of oil which assist in flotation, and they swim by characteristic jerky movements of their oar-like antennae and limbs. They are efficient collectors and consumers of phytoplankton, both as juveniles and as adults; in turn they are the principal food of the herring and of most of the other principal pelagic fish and invertebrates also. Ostracods or water fleas can also be very abundant, though in no way comparable in importance with the Copepods. The Euphausiids have already been mentioned, and their swarming in immense numbers. A few decades ago, one particular phytoplankton-eating species, *Euphausia superba*, supported a herd of about 800,000 whales in the Southern Ocean. Since intensive whaling has reduced these herds to a remnant, thoughts have turned to harvesting the food on which they fed, for direct human use, together with other abundant large members of the zooplankton such as the lantern fish and bristle mouths, the most abundant fish in the sea. Trials are being made, especially by Russian research vessels. Though commercially interesting catches have been made, they are not made regularly enough for commercial success.[9] Far more sensitive devices than the present radar and sonar will be needed for locating the shoals of zooplankton, devices rivalling those supplied by nature to the whales and fish.

There are local fisheries for the Euphausiid *Meganyctiphanes* in Japanese coastal waters, and for the mysid shrimp *Acetes* in estuaries and fishponds in the Indo-Pacific region. The latter may be fermented to make sauces and puddings.

Other Zooplankters

The various kinds of worms have members which have adapted to pelagic life, such as the annelid worm *Tomopteris*, which has its appendages enlarged and flattened for swimming. But the arrow worms or Chaetognaths are wholly pelagic. They are darting swimmers, by the movements of fin and tail-like outgrowths, and have a circle of sharp bristles round the mouth for seizing food organisms such as copepods.

Two groups of the molluscs are specialised members of the pelagic community, namely, the Pteropods and the Cephalopods. The Pteropods include the *Limacina retroversa* and *Clione limacina*, both of which are important foods for herrings, sprats and pilchards. *Limacina* looks like a pinhead-sized little black snail, and its presence is disliked by fishermen because a fish filled with this food does not keep well, and is said to have 'black gut.'

The Tunicates include many pelagic forms, and at least one of these, *Oikopleura*, can be an important item in the food of herring and young plaice.

Zooplankton-Eating Fish and Whales

Small pelagic fish, including not only the Scopelids but also the Cyclothones, the so-called bristle mouths, seldom exceed 60 mm. in length; but their importance lies in their abundance. They are probably the most numerous fish in the sea, occurring over the deep oceans as well as over the continental slopes. They feed on small pelagic animals including filter feeders, near the surface at night. By day, they migrate into deep water, as deep as 700 to 1,000 metres, and are an important means of carrying down to within reach of deepsea fishes the abundant food produced in the surface, photic, layers of the sea. They can be so plentiful that their shoals reflect the impulses of the echo sounder, and form the so-called 'deep scattering layer' which for a time mystified marine observers.

It was known that this layer consisted of living plankton because it moved towards the surface at night, and down again at dawn. One such layer was photographed from a submersible, and was identified as a dense mixture of small Scopelid fish, small squids, and the small fish *Sardinella*.[10]

The mackerels are important zooplankton feeders, and though not very important commercially in Europe, the *Scomber kanagurta* is very important in the Indo-Pacific area. But by far the most important zooplankton feeding fish are the clupeoids.

These zooplankton feeders have access to ten per cent of the primary production of the ocean (p. 145); it is no wonder, therefore, that zooplankton-feeding fish provide over forty per cent of the world's total fish harvest.

The cod, though usually a bottom-dwelling fish, at times lives pelagically over deep water, and another northern subarctic fish, *Sebastes*, which is now an important food fish, lives regularly there. A member of the cod family, the *Gadus poutassou*, well named the blue whiting by the trawlermen, is a pelagic fish, feeding chiefly on the Euphausian *Meganyctiphanes*, but also on small Scopelid fish, such as the luminous *Maurolicus*, which feed chiefly on copepods.[11] Trials are being made to develop the blue whiting into a commercial fishery, no doubt for fish meal manufacture, using midwater trawls, or some other pelagic type of net (Fig. 10).

The great abundance of the zooplankton, as the second step down from primary production, feeds not only the baleen whales, the largest animals that have ever lived, but also other very large animals such as the giant ray, *Manta*, or devil fish, which may weight two tons, and measure seven metres across the body, and the huge whale- and basking-sharks – all very harmless animals.

Active Members of the Pelagic Zone

Flying fish are among the best known of the pelagic fish, able to take off into the wind by rapid sculling movements of the tail, the lower lobe of which is elongated so that it continues to give an impulse when the rest of the body is airborne. As airspeed builds up, the wings, which are the very much enlarged pectoral fins, are spread, and the fish can travel several hundreds of yards before either re-entering the water with an untidy splash, or lowering its tail for another take off. This is the typical flight reaction to escape predatory fish.

Another oceanic pelagic fish is the *Coryphaena*, or dolphin fish, a beautifully-coloured and edible fish which has the habit of sheltering under floating objects; this habit is used for its capture. Artificial rafts are anchored in deep water, and when a shoal of coryphaenas has collected beneath one of these, a pursing net is shot around the float, and the coryphaenas caught. They are often accompanied by pilot fish, *Naucrates ductor*, more usually associated with sharks, which they are said to guide and with which they undoubtedly associate. Many kinds of shark wander about the oceans, and can be included among pelagic fish.

The squids and cuttlefish are also specialised for a pelagic life, and some live on zooplankton. I found them feeding on *Meganyctiphanes* on fishing grounds to the west of the British Isles, and they were in turn they prey of the predatory fish *Merluccius*, the hake, which was our quarry. Many squids live at great depths in the ocean, specialised for deep sea life. They have two modes of propulsion, a slow one, by the sculling of the fin-like planes on the body; and a very swift one, by jet propulsion. A very muscular mantle full of water discharges through a tubular siphon. This siphon can be turned in any direction; when the animal wants to travel fast, it compresses the muscles of the mantle to force a powerful jet of water from the siphon. These muscles can produce these jets with great rapidity, and speeds of 790 cm./sec. have been reported.[12] This is equivalent to fifteen miles per hour; these must be the swiftest of the marine invertebrates. They can also make considerable leaps out of the water, and often, at sea, a scatter of dead cuttlefish or squids, lying like collapsed paper darts, will be found in the morning on deck under the lights. In the deep waters of the oceans, squids are the chief food of the sperm whales, and some very large specimens have been found in their stomachs. One species, *Architeuthis dux*, may have a length of ten metres, and fragments of tentacles have been found which point to much larger squids than this.

The sperm whales feed on squids in oceanic waters, but over the continental shelves may feed on bottom fish. Off the coast of Iceland, bull sperm whales were found to be feeding on bottom fishes such as angler fish, lumpsuckers and sea perch;[13] off New Zealand they were feeding on lings and groupers.[14] A sperm whale bull, weighing many tons, would need to take a lot of fish, and may be an important factor in the ecology of such areas.

Crab-eater seals and penguins in the Antarctic feed on *Euphausia*; seals nearer home are troublesome to the salmon fishermen, and are having to be thinned out because they are breeding too fast. As they favour certain beaches to lie on, rejecting others apparently equally favourable, the increased seal population is very congested, and a high proportion of the pups are rolled upon and killed.[15]

The pelagic environment also includes, besides the whales, seals, sea otters and seabirds. All these except the whales must return to the land to breed. The baleen whales feed mainly on zooplankton and especially on the Euphausiids, though they will take fish, for it was common to be accompanied by 'blowers' when fishing herring, for example.

Since the rise of the great fisheries, many seabirds, notably the fulmar petrel, have become scavengers, following fishing boats and feeding on fish which have escaped through the meshes of the nets. In consequence of this abundant source of food, fulmars, and also gannets, have increased in number, opening up new colonies over a wide area.

The Sardines and Sardinellas

By far the most important pelagic fish of commerce are the members of the herring family and their allies. The oceanic tunas will be considered in Chapter 15. The Peruvian Anchoveta, *Engraulis ringens*, has already been mentioned, in view of its importance on the world scene. Another anchovy, *E. mordax*, has become the mainstay of the Californian industrial fishery since the collapse of the great Californian pilchard or sardine fishery. In the year 1935-6 the pilchard fishery of California gave a harvest of 560,000 tons, but by 1950-1 had declined to 353,088, and finally collapsed to 28,878 tons by 1960-1.

This dramatic collapse of a great natural resource was ascribed to selective fishing, reinforced by a succession of poor recruitment years. The reason for the failure of a season's spawning is but little understood in any fish. In the case of the Californian pilchard or sardine, some shift of the oceanic current system brought about unusually cold conditions which favoured the survival of the Californian anchovy, *Engraulis mordax*. Since this fish is in competition with the sardine for food, the advantage gained by the anchovy was consolidated by its taking over of the food supply. Now the sardine is a protected species, and the biomass of the anchovy has increased until it has replaced that formerly possessed by the sardine, and is a mainstay of the American fish meal industry.[16]

The Cornish pilchard fishery has a long history, and statistics of exports are available for over 200 years. The smaller and younger pilchards are caught as sardines, fish less than three years old, off the French coasts of Brittany; the Cornish fish are the same species, *Sardina pilchardus*, but from three years of age and older. The sardines are of course canned and provide an important industry for canned fish and fish meal all down the western coasts of Europe and Morocco, and in the Mediterranean. In the great days of the Cornish pilchard fishery, the fish were salted and pressed into barrels, the extracted oil having a good sale to the leather trade. The pilchard-curing cellars are still a feature of the older quayside houses, from Mevagissey to St Ives. There was also an important pilchard fishery in southern Ireland.

There has never been a sale for fresh pilchards in Britain, and when they are plentiful in the North Sea, as often happens, the herring fishermen regard them as a nuisance. Attempts to can them have failed, and at the time of writing the fishery is at an end for want of demand. It is certainly not due to a want of fish, for a survey of the western English Channel made in 1953 revealed shoals containing not less than 1,000 million mature fish.[17]

The South African sardine, *Sardinops ocellata*, swarms off the desert coasts of southwest Africa, feeding and fattening on the abundant plankton resulting from the upwelling of the rich deep water of the Agulhas Current. Like the anchovies of Peru, these sardines also support a large bird population as well as a great fishery for the fish meal industry. And as in California, the catch of sardines is declining while that of its competitor, the anchovy *Engraulis capensis*, is increasing.

The Shads

The Menhaden, *Brevoortia tyrannus*, has taken over from the Californian sardine as the chief source of fish meal and oil in the United States; and in fact provides forty-four per cent of all the fish landed there. None of the Menhaden is used for human food; they are small oily fish, mostly less than twelve inches (about 31.5 cm.) long, and in a typical year provide about 445 million pounds of fish meal, 20 million pounds of fish solubles, and 20 million gallons of oil. They feed on plankton, and mainly phytoplankton, and serve to emphasise the importance of the herring family in the world economy: for, without any wasteful intermediate steps, they convert the plankton into useful flesh and oil. But production has declined since 1962.

As stated in Chapter 5, the alewife has colonised the freshwater Great Lakes of North America; and the bonga, *Ethmalosa fimbriata*, plays a great part in the economy of West Africa, from the Senegal River to the Cameroons. They are caught on a large scale, and are phytoplankton feeders.

The Herring

The most celebrated and important of the pelagic fisheries (aside from that for the Anchoveta) is the herring fishery. For many centuries, the herring fisheries of the Baltic and North Sea supplied western and central Europe with its cheapest protein food. In the 1300s, the Abbott of St Albans was so keen to guarantee his supply of barrelled herrings that he opened his own curing houses in Great Yarmouth. The herring fishery gave wealth to persons and states, it founded the commercial empire of the Hansa (to this day the full style of Hamburg is 'the free and Hanseatic City of Hamburg'); and was one cause of the Anglo-Dutch wars of the seventeenth century.

The herring is a northern fish, occurring at Spitzbergen, Greenland,

Iceland and Labrador; but not in commercial quantities south of the Bay of Biscay on the European, nor south of Cape Hatteras on the American side of the Atlantic. The herring fishery is much subject to short-term and long-term fluctuations, due to the varying fate of the herring populations.

Fluctuations in the Fishery

The short-term fluctuations are usually the consequence of the success or failure of the breeding seasons. While it may still be impossible to explain these fluctuations, it is becoming possible to forecast them, and this is useful, bearing in mind that a glut of herring can be as embarrassing as a shortage. Surveys of the spawning grounds, and later of the nursery grounds, can give some indication of the success or failure of the year's brood of young. This preliminary warning can be followed up at a later stage by taking a census of the commercial herring shoals. The scales, bones and otoliths of the herring can be read to give a reasonably reliable indication of the age of the fish, and therefore of the year of its birth. Large-scale sampling of the commercial shoals will show which year classes are dominant in them, and which are poorly represented. This information should tally with the estimate made at the time of spawning and of the newly-settled young. If the fish begin to affect the commercial fishery in their fourth year of life, and even more so in the fifth and sixth and older years, then, knowing the relative strengths of these broods, it is possible to make a forecast of the contribution of each year class to the fishery, and thence of the contribution of all the year classes together. For many years past, forecasts of the likely results of the fishery have been made, helpful to the industry. It is even possible, from the abundance and distribution of the zooplankton, and especially of the copepod *Calanus*, the most important food of herring, to forecast the condition, and likely places of shoaling, of the herring.

Drift Netting

Until the Second World War, most of the herring were caught in drift nets, miles of light cotton (now nylon) nets hanging from floats near the surface of the sea, set at dusk. Especially near dusk and dawn, the herring rise from the deeper water where they have spent the daylight hours, and come near to the surface for 'the swim,' a state of activity in which they are especially vulnerable to the meshes of the drift nets in which they noose and strangle themselves. The hauling of the nets was a long and tedious job; for though a steam capstan took over the main load of hauling, the nets had to be shaken to release the herrings. When

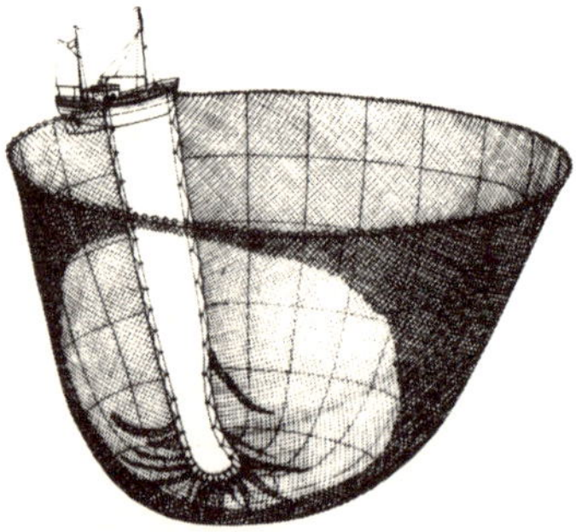

purse seine

drift netting

midwater trawl

Figure 10

the haul was a good one, the crew had to spend many hours in their yellow oilskins shaking and shaking, while scales and spray flew away downwind.

But there are also trawl fisheries for herring, in which the fish are caught while near the bottom, and these were increasing in importance, relative to the drift net, between the wars. But the herring fishery, as chiefly carried on by drift nets, stood up to heavy fishing at a time when the great trawl fisheries for such fish as cod, haddock and flatfish were in decline due to overfishing. It was thought that the selective action of the drift net, which allowed small fish to pass through unharmed, safeguarded the pelagic fisheries. But experience early in the last war showed that the large peacetime drifter fleets had in fact been taking a significant proportion of the herring shoals, for the reduced fleets due to the war took a much higher catch per boat than in peacetime.[18]

Modern Fishing Pressure

Since the 1939-45 War, the development of the echo sounder and sonar, which have greatly refined the process of locating fish shoals, in combination with more completely mechanised and unselective fishing methods, such as the purse seine and the midwater trawl, have increased the catch of herring (Fig. 10). The drift net now catches only a small proportion. Since the war, and especially in these last years, a very big industrial fishery for herring and mackerel has been developed by the Norwegians in the northern North Sea. So powerful has fishing effort become that, at last, the offtake of herrings is greater than the herring shoals can sustain, and massive overfishing is reducing the stocks.

Herring Races

In earlier days, it was thought that there was one great herring shoal, which appeared off the northeast of Scotland in spring, and then worked its way south during the summer, until in autumn and winter they were in the southern North Sea and the Dover Strait; and all along, the fish were subject to fisheries which moved south with the shoals. But in fact there are many races of herrings, which appear to move in a regular manner because the conditions herrings need for their breeding tend to move in a regular manner with the seasons. There are many small but locally important herring fisheries which are largely self-contained; but the most important fisheries are for the offshore or oceanic races which may make migrations of considerable magnitude.

In the North Sea there are three main races, the Scottish northeast coast or Buchan spawners, the Dogger Bank spawners, and the Downs spawners. The young of these three races mix on the same nursery grounds, in the central North Sea east of the Dogger Bank, in their second year of life, and further north in their third year of life. To the same grounds go the spent mature fish to recuperate. So fish which have spawned in the Dover Strait, the Downs herrings, may migrate long distances between spawnings.[19] The greatest of all the herring fisheries, and in its day one of the greatest fisheries in the world, was the autumn East Anglian fishery, based on Great Yarmouth and Lowestoft. It was worth a long journey to see this fishery at its height in October and November. But in 1947 the Danish fishermen located the one- and two-year old herrings in the eastern North Sea, and began a fishery which, by 1956, was taking 63,000 metric tons annually, a quantity which included at least 800 million undersized and immature herrings. These were rendered down to fish meal for the Danish poultry and dairy industries. Simultaneously, the East Anglian herring fishery for the adults fell from 45,764 metric tons in 1954 to 20,370 in 1955, and 8,550 in 1959. The greatest of all the herring fisheries was extinct by 1968. Though other explanations have been given the connection between the rise of the fishery for the baby herrings and the decline in that for the grown fish is convincingly close. The late Dr William Hodgson, one of the best of the herring experts, believed in this connection, and wrote that the herring shoals, which had for centuries fed Europe, were now being called upon to feed Europe's livestock also; and there was no proof that this could be done. The proof is still wanting.[20]

Herring Migrations

There are two recognised oceanic herring races, namely, the Hiberno-Caledonian, which inhabits and migrates within the Atlantic bounded by western Ireland and west and northwest Scotland; and the Atlanto-Scandian, which ranges over the Atlantic round Iceland, Jan Mayen, Faroe Islands, Shetland, the extreme northern North Sea, Norway and the north coast of Russia. Marking or tagging of individual fish belonging to this race have shown large migrations. Generally speaking, the Atlanto-Scandian race of oceanic herrings spend the winter to the south of their range, between the Faroe Islands and Iceland; but, in spring and summer, spread in a northerly and northeasterly direction as far north as Jan Mayen, Spitzbergen, and the edge of the ice, feeding on zooplankton. Fish tagged on the spawning grounds off the Norwegian coast have been recaptured in Iceland, and fish tagged in Iceland off Norway. Herrings caught while roving freely in summer in the Norwegian Sea have been recaptured travelling south to the wintering areas and east to the spawning areas.

But the climate of the Arctic is slowly changing. Over the last few decades the Arctic has been warm, but now it is beginning to cool; these cycles of temperature changes have long been known, and have an important effect on the fisheries. The Atlanto-Scandian herring stocks have been severely reduced since the mid-1960s; the Icelandic part of this stock has disappeared. The Norwegian spring spawners, which gave a fishery of 1.5 million tons in the season 1965-7, gave 20,000 tons in 1969-70. There has been a series of poor recruitment year classes, aggravated by a very high rate of exploitation of the immature fish, a rate estimated to be equal to, or even in excess of, the estimated total stock. Recruitment from more recent year classes of 1965-9 has been low, so no relief is in sight yet.[21]

The Former Plymouth Winter Herring Fishery

A similar change took place in the much less important winter herring fishery in the western English Channel, based on Plymouth; but this change was observed at the time and has an explanation, a tribute to the work of the laboratory of the Marine Biological Association at Plymouth.

In the period of warm years around the mid-1930s, the circulation of water in the North Sea was affected in such a way that the flow of water eastwards up the English Channel was reduced. Fish of northern origin, such as the herring, were replaced by fish of southern origin, such as pilchard. The winter values of dissolved phosphate decreased, leading to a reduced rate of primary production in the Channel, and a reduced quantity and quality of plankton. The recruitment of young herring stopped, so that in each successive season the older fish became fewer until the fishery collapsed.

But since about 1965, this process has gone into reverse. The winter values for dissolved phosphate have been rising, and with them the rate of primary production. Diatoms have become more abundant in spring and early summer than in earlier years, the copepod *Calanus* is once more very abundant in late spring, and the Euphausiid *Meganyctiphanes* has returned to the western English Channel after many years. The spawning of the pilchard has been reduced; but the young of many fish of northern origin, such as cod, haddock, and ling, have appeared, and are enriching the fish market at Plymouth. Most interesting of all, herrings are returning to the area, and a renewed fishery can be expected to develop in the next few years.[22]

And so these long-term climatic fluctuations swing to and fro, and men cannot affect them, only observe them. The herring fishery in the southern Baltic in the Middle Ages has long since vanished as a major fishery, and so have other herring fisheries. But the well-observed recovery to the earlier and more productive phase in the western English Channel must give hope to the victims of depressed herring fisheries.

NOTES

1. De Mendiola, B. Rojas. 1969. 'The food of the Peruvian Anchovy.' *J. Cons. Int. Exp. Mer.* 32 (3): 433-4.
2. Potts, G.R. 1969. 'Eruptive dispersal of shags due to food shortage.' *Inf. Inst. Mar. Peru*, 1969.
3. Longhurst, A.R. 1971. 'The clupeoid resources of tropical seas.' *Ocean. Mar. Biol.* 9: 349-85.
4. Wimpenny, R.S. 1966. *The Plankton of the Sea.* London, Faber & Faber, 1966.
5. Johannes, R.E. 1964. 'Phosphorus excretion and body-size in marine animals; microzooplankton and nutrient regeneration.' *Science* 146: 923-4.
6. Lohmann, H. 1903. 'Neue Untersuchungen über den Reichtum des Meeres an Plankton.' *Wiss. Meeresunters. Abt Kiel*, 8: 1-86.
7. Hobson, L.A., Mengel, D.W. and Barber, R.T. 1973. 'Primary productivity and pools or organic carbon in the mixed layer of the sea.' *Mar. Biol.* 19: 298-306.
8. Boltovsky, E. and Lena, H. 1970. 'On the decomposition of the protoplasm and the sinking rate of planktonic foraminifera.' *Int. Rev. Ges. Hydrobiol.* 55 (5): 797-804.
9. Ivanov, B.G. 1970. 'On the biology of the Antarctic Krill.' *Mar. Biol.* 7 (4): 340-51.
10. Schmidt, W. 1972. 'Deep Scattering Layers.' *Mar. Biol.* 16 (2).
11. Hickling, C.F. 1927. 'The Natural History of the Hake, Parts I and II.' *Fish. Invest. London. Ser 2*, (10) (2): 1-100.
12. Packard, A. 1963. 'Operational convergence between Cephalopods and Fish: an exercise in functional anatomy.' *Arch. Zool. Ital.* 51: 523-42.
13. Roe, R.J.S. 1969. 'Feeding habits of sperm whale at the Icelandic west coast.' *J. Cons. Int. Exp. Mer.* 33 (1): 93-103.
14. Gaskin, D.E. 1964. 'Recent observations in New Zealand on some aspects of behaviour of the Sperm Whale.' *Tuatara* 12 (2): 106-14.
15. Coulson, J.C. and Hickling, Grace. 1964. 'Breeding of seals.' *J. Anim. Ecol.* 33: 485-511.
16. Ahlstrom, E.H. and Radovitch, J. 'Management of the Pacific Sardine.' In *A Century of Fisheries in North America* Sp. Pubn. No. 7. American Fisheries Society.
17. Cushing D. 1957. 'The Number of Pilchards in the Channel.' *Fish. Invest. London Ser 2*, 21 (5): 1-26.
18. Hickling, C.F. 1946. 'The Recovery of a deep-sea Fishery.' *Fish Invest. London Ser 2*, 17 (1): 1-59.
19. Harden Jones, F.R. 1968. *Fish Migration*, Edward Arnold, London.
20. Hodgson, W.C. 1957. *The Herring and its Fishery*, Routledge & Kegan Paul, London.
21. Anon. 1972. 'Report of a working group on the Atlanto-Scandian Herring.' *Int. Coun. Exp. Mer.*
22. Russell, F.S., Southward, A.J., Boalch, G.T. and Butler, E.I. 1971. 'Changes in biological conditions in the English Channel during the last half-century.' *Nature* 234: 468-70.

Chapter 14 THE CONTINENTAL SHELF

The sea bottom slopes away more or less steeply from the tidal zone, over the Continental Shelf to the Continental Slope. Where the gentler contour of the Shelf gives way to the precipitous Continental Slope is the true boundary of the continent: the Continental Shelf is regarded as a continuation under the sea of the adjacent land. The Shelf generally ends, and the Slope begins, at about 200 metres, but off many coasts the shelf ends as 120-150 metres depth. The table on page 179 shows that the Continental Shelf occupies only about 7.6 per cent of the area of the sea bottom, and the Slope even less, at about 4.3 per cent.

The Continental Shelf is very wide along all the Arctic coasts of Eurasia and North America, and also off western Europe and the opposite shore of eastern North America. The Continental Shelf is formed partly by the debris of the land, material worn down by rivers and waves; but it may also be the result of earth movements. Almost within historical times, the North Sea and the Baltic were dry land, and the important Dogger Bank a range of low hills.

The Continental Shelf is an interesting part of the sea politically, because it is being increasingly claimed by riparian states as an underwater extension of its territory; industrially, because of the presence on or beneath it of valuable mineral deposits, including oil; and commercially, because they are the site of the world's greatest demersal fisheries. These demersal fish are almost entirely predatory fish. The political side is in flux, and a conference is being called to try to define and agree territorial waters.

The Continental Slope

Long before the last World War, English trawlers were fishing down the Slope as deep as 380 fathoms (about 750 metres) and the longline fishery for the Pacific halibut fished baited longlines as deep or deeper. Recently, the Soviets have pioneered a trawl fishery off Labrador in 1,000 metres, for halibut and grenadier fish.

But generally speaking, the Slope is an unimportant source of fish, and most of those caught have the wrong shape for human food, with their big heads and long thin bodies. In 1927, I was a member of a trawling expedition which worked as deep as 1,200 metres; the catches were only moderate, and the amount of edible flesh on the fish small.[1] Fried fillets tasted no different from those from better-known fish. If fisheries are developed in such deep water, they are likely to be used for fish meal.

Demersal Fisheries

The situation is very different on the Continental shelves. There are caught the most acceptable and valuable of sea fish, such as turbot, cod, brill, halibut, plaice, soles, haddock and hake. The rate of primary production is high over the shelves (p. 141), and this is reflected in high fish production.

The demersal, or bottom-living fish, are mostly caught in the trawl, a large bag of netting dragged along the sea bottom by powerful vessels, some of which are large enough to process their catch as soon as caught, while others bring their catch ashore to market. In shallower waters, the Danish seine is more successful, especially for the prime fish.

The sea bottom may be rocky and hard, or muddy and sandy. The muddy sea bottoms are generally the most productive. Formerly, rough bottoms, especially where melting glaciers had in the past dropped stones and boulders, or where rock outcropped, were fished by long-lines, miles of stout lines with hooks on short snoods at intervals of a few metres. The fish caught, chiefly halibut, with some cod and ling, fetched premium prices because of their good condition; whereas trawl-caught fish must undergo a considerable buffeting in the bag of the trawl. Steady improvement in the techniques of trawling over many decades now enable trawlers to fish successfully on grounds considered too rough for dragged nets. Slow growing deep sea corals and sponges, which originally cumbered the bottom, have long since been destroyed by repeated trawling. In past years, tons of these destructive corals would be heaped on deck amid the wreck of the net, and the sponges were huge pincushions of needle-sharp and brittle glass spicules.

The sea bottom on the Continental Shelf is generally too deep, and the water too turbid, for the growth of plants; so that the primary production which starts off the food chain is that of the phytoplankton, in the surface lighted zone. The new material produced is carried down to the bottom by the young stages of animals which when adult are bottom-dwellers, by fish and crustacea and other invertebrates which migrate between the surface and the bottom, and by the rain of detritus derived from the dead bodies of plants and animals of the plankton and detritus derived from the tidal zone.

I remember seeing the food chain well revealed in the catch of a steam trawler working in 300 fathoms off the southwest of Ireland. The predatory fish, hake, *Merlucius merluccius L.*, our quarry, was feeding on the deep water squid, *Ommatostrephes*, well-named the fantail by the fishermen. One of these squids was opened, and had its stomach full of the Euphausiid shrimp *Meganyctiphanes norvegica*; and when some of these Euphausiids were opened, their stomachs were full of a greenish mixture of phytoplankton and detritus.

A Medieval Iceland Cod Fishery

The commercial exploitation of the fishes of the Continental Shelf is nothing new. In the Middle Ages, fleets of fishing boats sailed in convoy from the now extinct port of Dunwich, and from Harwich, Blakeney and King's Lynn, to the Icelandic cod fisheries. A company in King's Lynn was formed to finance these ventures, and the Fishmongers' Company of London, then a power in the land, also financed many. These would be linefishing outfits, and the cod were split and dried down to stockfish or Poore John, a very merchantable article in the mediaeval and later markets. The Prior of Worcester regularly made large purchases of dried cod in Stourbridge Fair, sending wains to collect them.[2] It was one of the most easily available sources of protein food for use in the winter.

The Rise of a Trawling Industry

Nearer home, fisheries would be by hand methods such as lines and simple nets, from boats propelled by oar or sail. The trawl was well known under another name as early, certainly, as the Middle Ages. No doubt it would have been operated as the junk trawlers operate it from Hong Kong today, using the tides and currents to drag both boat and net along, and using oar and sail to work uptide for the next drag.

As time went by, and communications improved, the demand for fish would have risen to the point where larger vessels could go further offshore, where catches would have been higher. The use of ice to preserve the fish would have helped. But it is a remarkable fact that, in the Middle Ages, fishwives were able to keep prime sea fish alive all the way from the coast to wealthy inland markets such as Coventry, and get luxury prices. Even in these days of fast lorries, this would be none too easy a feat; by slow packhorse it suggests high skill.[3]

In the 1880s, steam began to supplant sail, and comparisons showed that the steamer was four times as effective as the sailer; now nothing could check the very rapid expansion of the fishing fleets, and the opening of ever more distant fishing grounds, until British fishing vessels covered an area from Morocco in the south to Spitzbergen, Iceland and Greenland in the north, and west to Labrador and the Newfoundland Banks. All the time, improvements were being made to the trawl, and larger and larger vessels were being built. Aids to fishing such as radio, radar, sonar and echo sounders were developed and improved.

At last, by the mid-1930s, there were no more fishing grounds left to explore, and the catch of fish first stabilised, and then began to decline.

The North Pacific Halibut Fishery

A similar course of events took place in the well-researched halibut fishery of the north Pacific. Though Captain Cook saw a subsistence halibut fishery there in 1770, a commercial fishery began in about 1890. At first, the fishing fleet increased in size and power, and the catch increased correspondingly. The nearer grounds soon began to give reduced catches to the long liners; but there were plenty of new grounds to open up further to the west and north. Soon the fishery had extended all the way from California to Alaska and the Bering Sea. The only two states concerned were Canada and the United States, and, as even the most distant halibut grounds began to be played out, both governments became worried. Soon after World War 1, a series of agreements based on a Halibut Treaty began the management of this international fishery, the first such managed fishery in the world. A programme of research was started; and, since the obvious reason for the decline was the taking of too many halibut, a reduction of the fishing effort was attempted, by imposing a winter close season in 1923, which proved ineffective. The method successfully chosen was a limit on the offtake of halibut, with a closure of the market when this limit had been reached. It was imposed in 1930, taking advantage of the great depression of that year which much reduced the demand for halibut, and laid up a large proportion of the halibut fishing fleet. These restrictive measures were resented by the fishing interests, and evaded as far as possible. But as the only markets for the halibut were in Canada and the northwestern United States, it was possible to keep a watch on infringements. To ensure the effective selection by the fishery of the well-grown halibut, the long line was the only fishing gear allowed; trawling, a nonselective fishing method which would catch very small and young fish, was prohibited on the halibut grounds, and a minimum size limit was imposed. The last measure was effective because it became unprofitable to fish on grounds where undersized fish were abundant. The length of long line and the number of shots made with it, were also subject to control (Fig. 11).

A Managed Fishery

The stocks of halibut responded at once to this reduction in the fishing effort, and began to build up once more.[4] The catch per unit fishing effort rose until the fishery was profitable again. This attracted more ships to the fishery, especially salmon fishing boats which were preparing for the summer fishing season. The stocks of halibut, as judged by the catch per unit of fishing effort, increased threefold since the regulations were imposed. As the stocks have built up over the decades, the allowable offtake has been raised, until the stage had been reached where an imposed decrease of thirty-five per cent in the fishing effort resulted in a sixty-five per cent increase in the permitted offtake. Because the reduced fishing effort gave each fish a better expectation of life, the proportion of 'large' halibut, fish of 60 lb. and over, increased tenfold. At a total cost of five million dollars, for the research and regulations, the value of the fishery has increased by 100 million. The maximum sustained crop of about 69 million pounds, makes the halibut fifth or sixth in value among the commercial fishes of the USA.

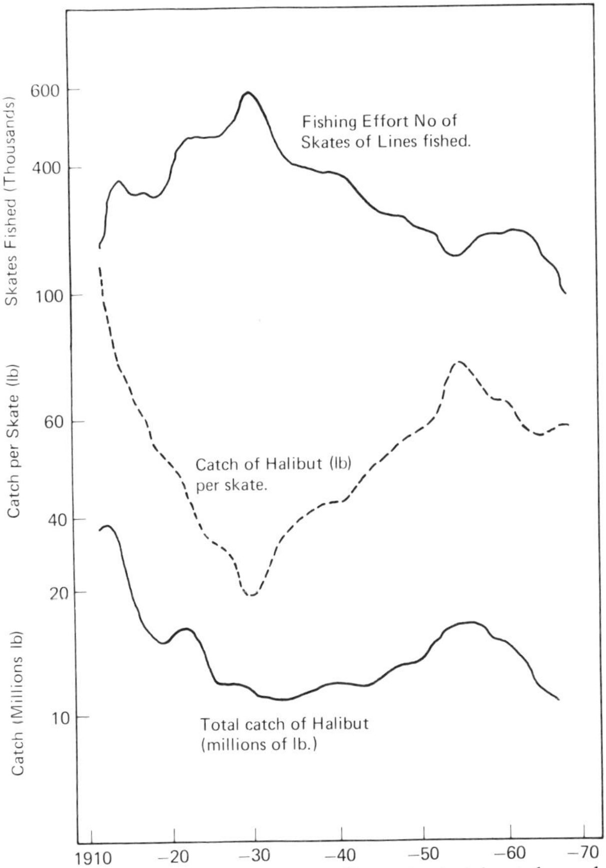

Figure 11 Relation between the fishing effort, the total catch, and the catch per unit of fishing effort, in the Pacific Halibut fishery in Area 2, south of Cape Spencer. From Heward Bell.[4]

Critics of the regulations said that in curing the inefficiency of unprofitable fishing, there had been substituted another inefficiency, namely, the inefficiency of a fishing fleet idled when the progressively shorter open season ended.

A Setback

In more recent years, the successful management has been upset by the entry of Japanese and Russian fishing vessels into the Bering Sea trawl fisheries, in addition to the Canadian and American. So long as the halibut fishery was a two nation affair and managed for their mutual benefit, success was achieved; but the Bering Sea is international water, and the Halibut Treaty could not be imposed on other nations without their accession to the treaty. The catches of these trawlers include large numbers of small immature halibut, taken in the course of fishing for other species; and though no doubt fishermen will return to the sea as many living fish as the working of the ship allows, many or most of these undersized halibut must die. The effect has been widely felt; for tagging experiments have shown that a large proportion of the bigger and older fish which sustained the fishery were raised in the Bering Sea, migrating eastwards as they grew, to replenish the halibut grounds south of Alaska and to the northwest coast of North America. The loss of these recruits to the fishery was estimated at five to ten million pounds weight annually. Besides this, the Pacific Fisheries Commission in 1963 set a catch limit for halibut in the Bering Sea much in excess of what was considered prudent. A year later, the halibut stocks were hard hit, and many fishermen suffered losses. The subsequent closing of the Bering Sea to halibut fishing came too late, though recovery would follow in due course.

Overfishing

This story of the management of the halibut fishery in the Pacific showed that the major changes had happened, not by natural causes, but by man's predation as a fisherman. When that rate of predation was reduced, the fishery recovered. The problem of overfishing has long been recognised in Europe, and though the remedy, a reduction in the fishing effort, was regularly debated, no agreement had been reached except for such harmless and ineffective measures as some regulation of the permissible mesh size of the nets, and minimum size limits for the marketing of the more important species of fish.

But debate was not in question when the two World Wars drastically reduced the fishing effort of all the European nations. The belligerents requisitioned their biggest, most modern and most powerful trawlers for naval duties, leaving only a small and weak fishing fleet at work. Neutrals were handicapped by minefields and the general hazards of war. As a result, plain for all to see, there was a rapid and complete recovery of the overfished stocks. I will quote a series of data well-known to me, to explain and illustrate the effect of the two world wars.[5]

History of a Trawling Port

The fishing port of Milford Haven was, in its heyday, one of the most important British fishing ports. Its rise may be seen in the table below:–

Table 3 Port of Milford Haven

Year	Total gross tonnage of Trawlers entering Dock, '000 tons	Weight of Fish despatched, Tons	Weight of Fish despatched per 100 Trawler Tons
1890	230	9358	4.07
1895	250	11344	4.54
1900	449	21437	4.77
1905	623	31825	5.11
1910	728	39406	5.41
1912	720	34194	4.75
1916	341	19657	5.80
1918	263	18819	7.16
1922	950	46271	4.87
1925	1053	47739	4.53
1929	1051	45460	4.32
1930	1151	47540	4.13
1931	1232	51357	4.16
1932	1013	43981	4.34
1933	1075	44345	4.12
1934	1003	38228	3.81
1935	1005	38194	3.80
1936	990	41967	4.24
1937	1092	44879	4.11
1938	970	44118	4.53
1942	227	33038	14.55
1944	201	31947	15.87

The table shows the rapid expansion of the fishing fleet based on Milford Haven from 1890 to 1912. The weight of fish despatched from the port also increased, in fact it increased a little faster than the increase of tonnage; for the newer ships were larger and more powerful than their predecessors, thus continually increasing the fishing effort at the port. The last column shows that the weight of fish despatched per 100 trawler tons based at the port increased from 4.07 in 1890 to 5.41 in 1910.

World War 1 and the Interwar Years

Then came World War 1. In 1916, the fishing effort, as measured by the gross tonnage at work, fell by half, but the fish increased to 5.80 tons per 100 trawler tons. In 1918, the tonnage fell still more, to 263,000 tons, or one third of 1912; yet the fish despatched were more than one half of the 1912 figure. Fewer ships were each catching more fish.

By 1922, the fleet had increased to 950,000 tons, a far higher figure than any prewar year. The new ships were large trawlers originally built for minesweeping and patrol duties, and they added greatly to the effectiveness of the fishing effort. The total weight of fish despatched was higher than in any prewar year; yet the catch per trawler-ton had fallen to little more than one half of that of the small fleet of 1918. A bigger fleet was catching less per ship per unit time. But even this could not be sustained. Though the size of the fleet continued to increase slightly, catches did not increase, and the catch (fish despatched) per unit fishing effort (gross tonnage) fell to 3.80 in 1935. There was a recovery from 1936 to 1938, associated with some reduction in fishing effort, but the catch per unit fishing effort was still far below that of the 1920s.

Then came World War 2. By 1942, the fishing fleet had shrunk to the level of 1890, and now consisted almost entirely of small and elderly ships, as they would need to be to escape naval requisition in those difficult days. Veterans forty-five years old were catching quantities such as they had never caught before. The catch per 100 tons of trawler tonnage had risen nearly fourfold. By 1944, an even smaller fleet caught nearly three-quarters as much fish as the big fleet of the 1930s.

These greatly improved results were due to two factors; the first was an increase of fish caught by each ship, because fewer ships were at work. Over a long period of years, it can be shown that the catch of fish per ship varies inversely with the size of the fleet at work. Such a relations could happen only if the fishing fleets were making significant inroads into the fish stocks. In the 1890s and 1900s, the small fishing effort of those days was not enough to affect the stocks seriously. But the high fishing effort of the 1920s and 1930s was indeed affecting the fish stocks.

The Effect of a Reduced Fishing Effort

But the main effect of the reduced fishing effort of the two wars was seen in 1918 and 1942, when there was a great increase in the proportion of big fish, as well as in their abundance. As with the Pacific halibut, the reduction of the fishing effort had increased the survival rate of the fish stocks; more survived to be two years older, during which time they double their weight, and then double it again. The overfishing of the interwar years had so reduced the survival rate (to thirty per cent per annum or less) that fewer and fewer reached large size, and the fishery depended on the intake of small fish. The proportion of large fish on the market fell, and that of small fish, derisively called 'needles' by the trade, rose. This was corrected by the reduction of fishing effort by the wars. The high proportion of big fish now gave much bigger catches for the same fishing effort. War did what no reasoning could do; it seems almost impossible to concert the fishing effort of so many nations in peacetime – even the American halibut management broke down when other nations entered the fishery.

Limits on Fish Production

After 1942, the stock of fish, as judged from the fishing results, did not increase much further. Not only do predatory fish tend to feed on the smaller fish of their own species, as well as on other species of fish and squid, but the absolute amount of fish and other predators which a fishing area can carry is limited, in the final resort, by its rate of primary production. For the seas as a whole, it was shown, in Chapter 12, that the primary production could be estimated at about 20,000 million tons of carbon. This would be reduced down the food chain to about 20 million tons in predatory fish and squids. But since fish contain about ten per cent of carbon in their tissues, a total of about 200 million tons of predators, chiefly fish, can be supported. But not all of this is fish; not all predatory fish are marketable nor catchable under present conditions. And as fish stocks are thinned out by fishing, it may become uneconomical commercially to fish them, even though the fish are still plentiful. If this 200 million hypothetical tons of predators is reduced by a factor of five to allow for these and other adverse factors, as has been suggested, then the weight of predatory fish available for capture in all the world's seas is about 40 million tons. The actual figure of world catches is well within an order of magnitude of this; mankind may be within sight of the practical limit of the predatory fish the seas of the world will give him. In Chapter 13, the possibility of catching fish and plankton higher up the food chain was mentioned. There is almost unlimited scope for expansion at those levels, if a practical method can be found. The product would be some kind of high protein meal; and though tests have shown a poor appreciation of this kind of product, the blasts of the public relations men could change all that in a skilled promotion campaign.

The North Sea

The history of the North Sea has followed a different course. From 1912 to 1917, the catch of all bottom-living fish fell from about 450,000 tons per annum to 200,000. Then came recovery to 550,000 tons in 1920. But from that year until 1938, the catches trended downwards to about 350,000 tons. The proportion of small fish in all species rose continually except where an exceptionally good year class gave a temporary boost. The effect of the Second World War was more drastic than the First, for naval and air activity was far more adverse to the fishing industry than in the First; the catches fell to little over 100,000 tons, one half of the lowest of the World War 1 years. Recovery followed from 1941 to 1945, for here also the reduction of fishing effort resulted in a great increase in the proportion of large fish, notably plaice.

It is a fact that there is no relation between the abundance of spawning fish and the success of the year class of young fish produced. The great increase in the number of big and mature fish resulting from the wars did not result in more successful year classes – nor, comfortingly, did the thin stocks of mature fish in the overfishing days result in poor year classes. The causes of the success or failure of year classes are still not well understood; but the abundance of spawners is not usually one of them. A given area of sea can raise only so many young fish; when this number has been spawned, further additions are ineffective.

Since 1950, the catch of fish from the North Sea has increased very much, to nearly 700,000 tons, as compared with the previous best of 550,000 tons in 1920. New techniques and some successful year classes account for this, beside an intensive industrial fishery for herring and mackerel, using asdic and big purse seines (Fig. 10). But the recent very high rates of fish catches in the North Sea are considered to be near the limit of what can be grown there, and are not expected to continue for long at that level.[6]

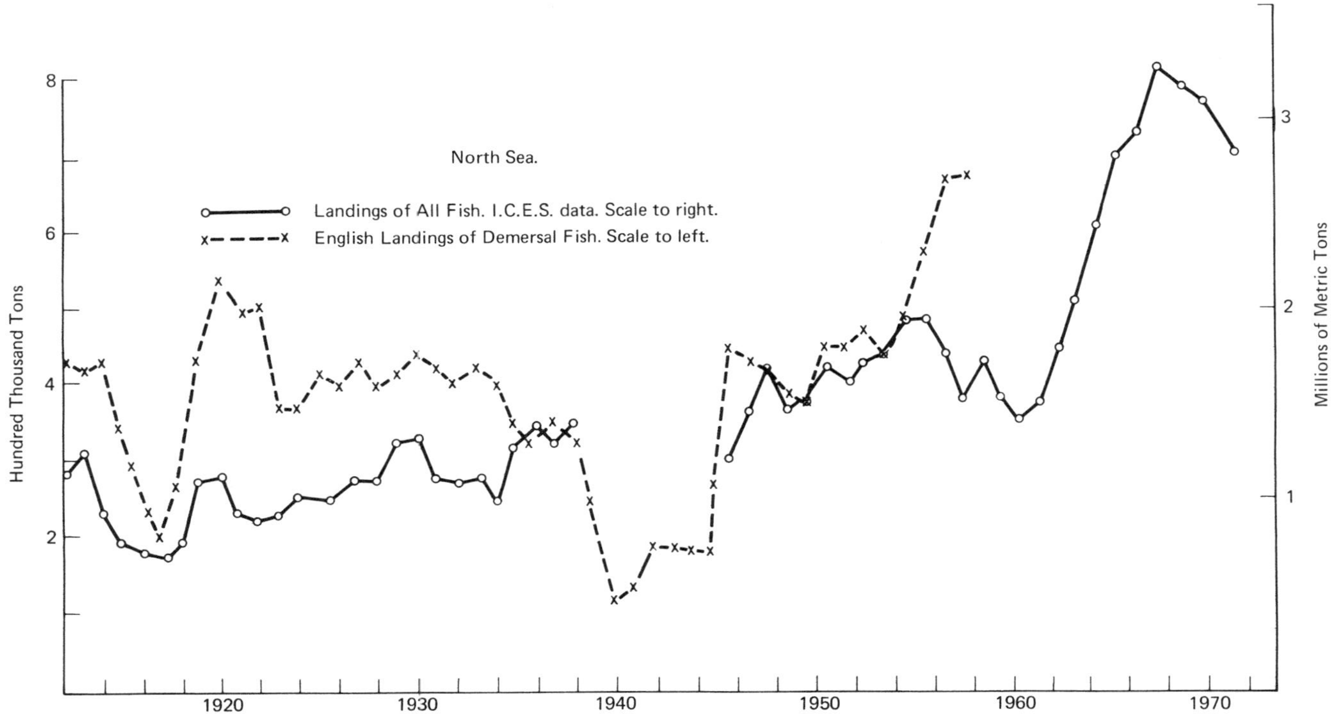

Figure 12   Catch of fish from the North Sea, 1912-71. English data for demersal fish only, international data for all species of fish.

Fish Migrations

Fish migrate as they grow, and so change their habitats and their value as raw material for the fishing industry. The most obvious migrations are for the purpose of spawning. Just as, in a stream or river, fish must swim upstream to lay their eggs, so that the growing young can be carried back to the area they will inhabit when adult; so, in the sea, spawning fish swim upcurrent and spawn where the young, when they hatch out, will be drifted back to the preferred growing areas. Within these areas, there are feeding migrations; an example is the migration of the cod in pursuit of the herring and capelin.

These spawning migrations do not always result in a successful spawning, as has been said. The current systems, for instance, can alter according to meteorological conditions which determine the strength and direction of the winds which drive them. Some fish which have been studied seem to have brood successes correlated with climatic conditions, and successful predictions have been based on this. If the brood of young fish is drifted by unusual current conditions over deep water or to unfavourable feeding areas, or if temperature conditions are unsuitable, a poor brood-year may result. Or if unfavourable conditions result in a shortage of zooplankton for the nourishment of the young fish, or if there is a glut of planktonic predators such as jelly fish, again the survival rate of the year's spawning may be poor, and will be reflected in poorer catches at the time when these fish would have grown to commercial size. There is, however, a slight compensation, for a poor year class, having more space and food than a good one, may grow faster and so enter the commercial catches sooner.

In extremely adverse conditions even the older fish may be killed, as in the severe winters of 1947 and 1963, when the fishermen caught thousands of dead soles in the North Sea.

Tagging of Fish

The migration of the most important species of fish are now becoming well-known, thanks to tagging experiments. As the name implies, tagging involves catching the fish alive and in strong condition, measuring and weighing them and determining if possible their sex, then placing a numbered tag of metal or plastic where it will be noticed, and may stay on the fish for years, and releasing it. On recapture many facts can be found about the minimum distance travelled, the rate of growth, and the rate of mortality. But there are snags which have to be allowed for. Plainly, tagged fish will not be recaptured where no fishing is taking place; and the larger tagged fish are much more likely to be spotted by fishermen as they lie on deck than the smaller ones, so a bias in favour of the larger fish has to be allowed for.

Plaice

As an example well-known to me personally; in the years 1929 to 1932, nearly 20,000 young plaice were tagged in the North Sea by an English research ship, on the young plaice grounds along the coasts of Holland, Germany and Denmark. Some 6,700 were recaptured and returned to us for the stipulated reward; though, amusingly, some fishermen refused a reward because they said the fish were ours since we had put our labels on them.

All the plaice tagged were small and immature fish less than 25 cm. in length, growing slowly on the usually overcrowded nursery grounds of the eastern North Sea. The tagging returns showed no definite migrations among these youngsters. They spread out from the point of tagging over the inshore waters like a drop of ink in a basin of water. As late as the third year after tagging, most fish were recaptured near the position of release. But the larger fish spread more widely than the smaller and grew more rapidly; the data did not indicate whether they grew faster because they travelled further, or left the nusery grounds because they were growing faster. Plaice from all coasts tended to mix in the slightly deeper water between the coasts and the Dogger Bank.[7]

The rate of mortality revealed by these experiments was high, namely, about seventy-eight per cent per annum off the Danish coast, seventy-one per cent off the Frisian coast, and sixty-seven per cent off the Dutch coast. High as this seems, it is about the same as earlier tagging experiemts in the years 1905 to 1913 had revealed.

As the plaice become mature, they do make definite migrations to the spawning grounds. Several such grounds are known around the British Isles, and of course many others are known in the Baltic, White Sea, Iceland and Faroes. The plaice is not found off Greenland nor Jan Mayen, nor along the eastern coast of North America.

Plaice Spawning Grounds

In the North Sea are four main plaice spawning grounds, namely, off the east coast of Scotland, off Flamborough Head in Yorkshire, in the Heligoland Bight, and, most important, in the Straits of Dover. It is not certain if the fish always return to the same spawning grounds, like salmon returning to the river of their birth; but certainly between the spawning seasons there is a great deal of mixing of fish from the various spawning grounds. An illustration can be given from the tagging experiments already mentioned: two plaice tagged off the Hook of Holland were recaptured within a few days of each other off Start Point in Devonshire. They had travelled a minimum distance of 350 miles, and had grown well in spite of their exertions. They were recaptured much nearer to a spawning ground off the north coast of Cornwall than to the Straits of Dover. Plaice spawned off the north coast of Cornwall are drifted by winds and currents north into the Irish Sea as far as the Isle of Man and the Cumbrian coast. Presumably, when they mature, they will migrate south to Cornwall again to spawn.

Icelandic Plaice

The Icelandic plaice have also been studied. They give a valuable fishery all round the Icelandic continental shelf, to a depth of eighty metres. The southwest coasts of Iceland receive warm water from the Atlantic Drift, while the north and east coasts receive cold currents from the Arctic. The Icelandic plaice, as they mature, gather for spawning in the warmer water off the southwest coast. Fish tagged on the cold northeast coast move round in both an easterly and a westerly direction to the warm southwest coast. As the main currents run in a clockwise direction around Iceland, some of these plaice will be migrating with the current, and some against it.

Iceland-Greenland Cod Migration

The cod is a much more powerful swimmer than the plaice, and makes greater migrations.[8] For centuries, this fish has been the mainstay of the European fisheries. It occurs on both sides of the north Atlantic as far south as Carolina and the Bay of Biscay respectively. It is fished on the Newfoundland Banks, Davis Strait, east and west Greenland, off Jan Mayen, between Norway and Spitzbergen to the edge of the pack ice, and east to Novaya Zemlya, and, of course, off Norway, in the North Sea and in the Baltic. The Icelandic cod fisheries have been exploited by English and other fishermen since medieval times, and a large-scale cod fishery at Newfoundland was recognised within three years of Columbus' crossing of the Atlantic; this suggests some earlier knowledge by Europeans of this very rich fishery.

There seem to be both local and oceanic cod stocks. Local stocks include those in the North Sea, the Baltic Sea, and the Norwegian and Greenland fjords. There are two main oceanic cod stocks, the Iceland-Greenland migration, and the Barents Sea-Norwegian migration. The stocks on the Newfoundland Banks seem to be selfcontained. But there is much mixing among these cod stocks.

Cod tagged on the spawning grounds to the south of Iceland have been recaptured on the Newfoundland Banks, on the west coast of Greenland, the Faroe Islands, northern Norway, and even in the North Sea. If some fish can make these 1,000-mile wanderings, many can, and such a mixing of the cod stocks can mitigate the effects of bad year classes. For example, a successful year class grown on the west Greenland grounds contributed a large proportion of the cod caught at Iceland in 1970, where there had been a poor year class.

In fact the southeast coast of Iceland is the most important spawning ground of the Iceland-Greenland spawning group of cod. The developing eggs spawned there are carried by the Irminger Current to the north and east coasts of Iceland, and also west to the wide continental shelf off the desolate east coast of Greenland, where new trawling grounds have recently been opened up. Thence the stream of developing young cod drift round Cape Farewell and north up the Davis Strait. Here, the young fish grow until they become mature in about their ninth year of life. But these west Greenland cod are not all derived from the Icelandic spawnings, for there are some spawning grounds off west Greenland too. There is a sibstantial return migration to Iceland to spawn, these fish contributing substantially to the crop of cod there.

Spawning Grounds of the Barents Sea-Norwegian Cod Migration

There are spawning grounds for the cod of the Barents Sea-Norwegian migration all along the coast of Norway and round Finnmark to the north coast of Russia. The chief spawning ground is the famous one near the Lofoten Islands and in the Westfjord. Here occurs one of the world's most intensive fisheries; forty million cod may be caught there in one season by thousands of fishermen using every method from hand lines to purse seines. Some of the spawning cod are from local fjord stocks, recognisable by distinctive features such as their shape, size at first maturity, number of vertebrae and rate of growth. But most of the spawners are from the Barents Sea-Norway migration. This group of cod grew up in the Barents Sea, spreading in summer over the wide area of the Continental Shelf, feeding vigorously. In the winter they withdraw south and southwest towards warmer and deeper water off Bear Island. This migration is a leisurely one, averaging three miles a day. In the spring, they return north; this annual cycle is repeated until the fish become mature at eight to ten years of age. In the year before maturation, which might be called the year of adolescence, the fish may make a 'dummy run' to the spawning grounds off the Norwegian coast, but this time at the higher average speed of seven miles a day.

At full maturity, these cod quit the Barents Sea and move south along the Norwegian coast to the spawning grounds. The mature fish do not necessarily spawn every year, but those which do make a fast return migration, as spent fish, to the feeding grounds in the Barents Sea, averaging eighteen miles a day. There they spend the summer, like the juveniles, feeding vigorously to replace the material lost in the spawning, and to continue their growth. Herring and capelin are the most important items of their food. Meanwhile, the eggs they had spawned off the coast of Norway drift north as well on the currents, developing into young which finally descend to the bottom of the Barents Sea, to repeat the life story of their parents.

Predictions

The abundance and distribution of the O-group cod, as the young of the year are called, are carefully surveyed and assessed with echo sounders and the catches of special small-meshed nets, by all the participating fishing nations, and on a scale large enough to give a reliable census. From their abundance or want of it, predictions can be made of future catches. For instance, the spawnings of the years 1965 to 1967 were failures, and few O-group cod were found. Predictably, the catches of small cod in 1970 were poor. But the year class 1970 was a successful one, so a good supply of small cod can be expected from 1975 onwards, and the survivors will boost the catches of large cod in 1979.

NOTES

1. Hickling, C.F. 1928. 'The Fleetwood Exploratory Voyages for Hake.' *J. Cons. Int. Exp. Mer.* 3 (1): 70-89.
2. Fegan, Ethel S. 1914. *Journal of Prior William More*. Worcestershire Historical Soc. London, Mitchell Hughes & Clarke.
3. Davis, Dorothy, 1966. *A History of Shopping*. London, Routledge & Paul Paul, 1966.
4. Bell, F. Heward. 1970. 'Management of the Pacific Halibut.' In *A Century of Fisheries in North America*. Sp. Pubn. No. 7. Amer. Fish. Soc. Washington D.C. 1970.
5. Hickling, C.F. 1946. 'The Recovery of a Deep-Sea Fishery.' *Fish. Invest. London, Ser II*, XVII (1) London, HMSO. 1946.
6. Gulland, J.A. 1968. 'Food-chain studies and some problems in world fisheries.' In *Marine Food Chains*. Edinburgh, Oliver and Boyd, 1970: 296-315.
7. Hickling, C.F. 1938. 'The English Plaice-marking experiments.' *Fish. Invest. London. Ser II*, XVI (1) London, HMSO. 1938.
8. Harden Jones, F.R. 1968. *Fish Migration*. London, Edward Arnold Ltd. 1968.

Chapter 15 THE OCEANIC ENVIRONMENT

The area of the Continental Shelf, dealt with in Chapter 14, and defined as the area of sea less than 200 metres in depth, occupies about 7.6 per cent of the area of the oceans. Below are given the percentages of sea bottom beneath each depth.

Depth, metres	Per cent of oceans and seas
0- 200	7.6
200-1,000	4.3
1,000-2,000	4.2
2,000-3,000	6.8
3,000-4,000	19.6
4,000-5,000	33.0
5,000-6,000	23.3
6,000-7,000	1.1
7,000 +	0.1

The Continental Slope, which is the steep scarp connecting the shallow Shelf with the ocean depths, occupies only 4.3 per cent. By far the greater part of the oceans is between 3,000 and 6,000 metres deep, comprising seventy-six per cent of the ocean area. Water deeper than 6,000 metres, including the very deep trenches, comprises only 1.2 per cent. The deepest known trench, the Mariana Trench, is 5,933 fathoms, or 10,950 metres, deep, and it is not likely that a deeper one will be found.

In the early days of oceanography, when the contours of the bottom of the ocean had to be deduced from soundings laboriously taken by sinking weighted lines, such soundings were few and widely spaced. It was thought that most of the ocean floor was flat or gently contoured. But with the perfection of the echo sounder, a continuous written record of the bottom profile beneath a ship at speed can be got. The closely contoured charts now available show that the profile of the ocean bottom is as rugged as the land. The fascinating studies of the volcanic ocean ridges, and of continental drift, belong to marine geology.

This vast area of ocean is mostly remote from any land, and not much affected by it, though, for example, at the foot of the Continental Slope near the land masses enough detritus of terrestrial origin, such as the breakdown products of the seaweed beds, settles to support a biomass of 1.5gm/m^2. Life exists at all levels in the ocean, even in the greatest depths known. The two brave observers who reached the bottom of the Mariana Trench saw two creatures, a prawn and a flat-fish, though the latter was more likely to have been a flat holothurian. The deep sea expedition of the Danish 'Galathaea' found unexpectedly large numbers of invertebrates in the mud in the deepest positions sampled. Naturally, life is most abundant where primary production is taking place, in the uppermost 100 metres, and becomes progressively sparser the further from the surface. But there seems to be some increase at the bottom, because all the material which falls from above must accumulate there. The source of this production is the phytoplankton, including the nanoplankton, except on coral reefs, where production is by the reef-forming algae and the symbiotic green algae living in the corals.

An Oceanic Food Chain

The organic matter newly-produced in the uppermost photic layers of the ocean must be carried deep down to nourish the abyssal fauna. In the ocean areas sampled by Hobson *et al*,[1] the microzooplankton which fed on the phytoplankton consisted chiefly of ciliates, and copepodite and nauplius stages of copepods, at all stations. This microzooplankton will be food for larger filter feeding animals. For long it was thought that the surface flora and fauna, living their brief lives in the surface waters, maintained a steady rain of dead bodies on which the deeper living organisms could feed. But in the ocean it is unlikely that any digestible organic matter could fall far before it would be consumed by another animal, unless it were very large, such as the body of a large shark or whale, or waterlogged fragments of wood.

From the faunistic point of view, the upper waters of the ocean can be divided into a thermosphere, where the eurythermal pelagic fauna lives in temperaures up to 25°C; and a psychrosphere, divided from the thermosphere by the arbitrary temperature of 10°C, where lives the cold-tolerating deep water fauna. This dividing line is found at depths varying from 500 to 800 metres according to the ocean.[2] The organisms of the thermosphere perform diurnal vertical migrations, especially the larger species such as the Euphausiids and fishes, coming up to the surface waters at night to feed on the material which has been produced by the phytoplankton during the hours of daylight, and descending to the depths by day. In mass, they give rise to the 'deep-scattering layer' detectable by echo sounder.

Ferguson Wood gives us a good example of this.[3] In the Gulf of Guinea, he found that green flagellates belonging to the Coccolithophores multiplied in the sunlit surface waters of the ocean, until they were to be numbered by millions per litre. But in the evening and early night, their numbers fell to only tens of thousands per litre. This reduction coincided with the upward vertical migration of the zooplankton.

Vertical Migrations

These vertical migrations may cover several hundred metres, a long trip for small organisms; and this would be worth while only if the better feeding conditions near the surface provided more energy than would be used in the process of migration. But it has been shown that only a small amount of energy is used in the ascent, and even less on the descent. Besides the availability of abundant food in the surface layer of the ocean, the migrating organisms gain several advantages. They spend the dark hours feeding in the warm near-surface water, up to 25°C in the tropics, yet by day descend to cool water of 10°C or thereabouts, where they can digest their meal at this temperature, which makes them use less energy and need less for floating in the denser water. The saving is substantial, and enables the organism to put by large stores of energy which is available for reproduction, or for the fish which prey on them.

At about 500 to 800 metres depth, the water becomes too cold for these warmth loving denizens of the thermosphere so that their downward migration ends where the water has a temperature of about 10°C, and the fauna tend to accumulate there. But there, also, is the upper limit to the movements of the cold-tolerant fauna of the psychrosphere, which are checked by the higher temperatures. The deep sea pelagic animals of the psychrosphere feed well on the layer of food organisms from the thermosphere. They include fish, prawns, arrow worms and squids. In their turn, these deep water psychrophilic animals are food for still deeper living fish, now predominantly members of the Ceratoideae, the deep sea angler fish. In this way the primary products of the daily phytoplankton activity in the surface layers of the ocean are carried deep and distributed.

Important Vertical Migrants

As an example, the small deep sea fish *Benthalbella infans* belongs to the Lantern fishes or Myctiphoidei. Its distribution is worldwide in all the oceans, and in the eastern Atlantic its distribution in depth is from 90 to 1,500 metres, a wide range. At night there is an upward vertical migration of adults and juveniles, and some rise to near the surface. The smallest stages, to 28 mm. in length, live at depths from 100 to 300 metres, and the older stages live progressively deeper; this is usual among fish.[4] The adults are hermaphrodite, probably because their opportunities for mating are less common than among most other fish. The larger specimens were feeding mostly on small fish, specimens of 7 to 12 cm. feeding on bristlemouths (*Cyclothone*) of 1.8 to 2 cm. length, and fragments of myctophids of about 4 cm. length. There were also an eye-lens and beak of a small squid in an 11 cm. fish. The very wide range of depth of this fish makes it a typical link between the upper photic layer of the ocean and the deeper ocean to 1,500 metres depth or more; for this small fish will itself be welcome food for larger fishes and other predators still deeper in the ocean.

There is another way in which the primary production of the ocean can be carried down to feed the very deep dwelling organisms. Many of the deep sea eels spawn eggs which develop into leptocephalus larvae. These migrate upwards to the photic zone, much as the younger stages of the *Benthalbella* were found nearest the surface. As they grow and metamorphose into small eels, they make the long migration down into the abyss from which their parents came. As they may occur in billions, they are welcome food for all the predatory animals among which they pass, a manna from heaven for so many hungry mouths, bringing the goodness of the sunlight into the sunless depths.

Economising Energy

These deep sea fishes have various ways of economising energy.[5] One group has no swim bladder but are at near neutral buoyancy because they have very soft watery bodies containing 88 to 95 per cent water. These fish include small *Gonostoma, Melanostoma* and *Chauliodus*, which also have little haemoglobin in their blood. They are capable of short bursts of activity during which they live anerobically, recovering slowly later. The second group have swimbladders and can compensate for buoyancy at different depths; so their bodies are firmer and contain about 70-83 per cent water, while their blood has a much higher haemoglobin content. Both these groups of deep sea fish perform active vertical migrations. The third group include the familiar flying fish and mackerel, with hard bodies containing 64 to 74 per cent water, a high haemoglobin content, and adapted to active surface swimming. There are many devices for achieving flotation without the expenditure of energy. The deep sea sharks have huge livers, as large as 25 per cent of the weight of the body, which contains a light oil called squalene, and this gives the fish a neutral buoyancy.[6] Many deep sea invertebrates such as squids achieve lightness, relative to the seawater they live in, by accumulating ammonium chloride in their tissues in place of the heavier sodium chloride. Cuttle fish secrete gas, chiefly nitrogen, in the pores of the 'bones' so well-known to seaside strollers, a considerable feat physiologically.

Coloration and Vision in Deep Sea Fish

Though the amount of original primary production must lessen the further down from the sea surface, fish are more plentiful than one might expect. The bristlemouths may spaced on average only three metres apart, and the ceratioid angler fish thirty metres apart at depths of 1,000 to 2,000 metres. The colouration of the deep sea pelagic animals is designed to give protection from enemies, and luminous organs are confined to such fish and are not found on fish which live on the ocean bottom. The fish which live near the surface by day are blue in colour, with silvery sides and a white belly. The blue back makes them invisible from above, the silvery sides reflect the surrounding colours and make the fish less visible from the side, and the white belly below a darker back has an obliterating effect. Fish living in weak light tend to be red, black, or brown. Deep sea prawns are red, for in weak light both red and black are obliterating colours. Luminous organs, when present, are arranged to confuse and deceive predatory fish. However, in spite of these devices, fish and other organisms are caught and devoured, or predators could not exist in the deep midocean.

In the deep water, the eyes of fish tend to become smaller though at the same time they are modified to make the best use of what light, daylight or animal luminescence, is available. Other organs of sense come to play a major part in the life of deep sea fish, such as the organs of the lateral line, which have long been known to be sensitive to low frequency disturbances in the water. There is probably some kind of sonar, and fish and other organisms make sounds, which must be detectable by others. An American worker has reported detecting a form of radio waves, called hydrionic radio waves, which are neither sonic nor electrical, but which can be received by submerged dipole antennae.[8] He found that each fish species has a characteristic carrier frequency, pulse duration and intensity. There were more of these signals by night than by day; predators send out more powerful 'search' signals than nonpredators; and fish with poorly developed or vestigial eyes may emit frequent and strong 'search' signals.

The Whales

The whales are probably the most spectacular of the ocean's inhabitants. They are now being hunted to danger point. At one time, the population of the baleen whales, those with no teeth but with sieves in their great mouth cavities to strain off the small organisms on which they feed, was estimated at 800,000 in the Antarctic alone, and they were consuming about 170 million tons of zooplankton, almost wholly

Euphausia, or 'Krill,' annually. As everyone knows, the greatest of these, the blue whale, has been shot out almost to extinction, and a residual herd, estimated at only a couple of thousand, is now protected. The other big whales such as the fin whale are also being ruthlessly thinned out; and attempts over the last few decades to limit the offtake of the herds of the big whales have had little success. Even the five years' respite from hunting during the Second World War failed to produce any recovery over the prewar figures. Unlike fish, whales are slow breeders. The southern fin whale does not breed until the sixth to seventh year in both sexes, and then produces only one young annually, often omitting a year. In the last four decades, the age of maturity has lessened; in the years 1910 to 1930, when the whale herds were large, maturity in the fin whale set in at the age of ten years. But since about 1930, the weight of whale hunting switched from the vanishing blue whale to the fin whale, and the fin whales themselves have been thinned out to the point where there is more food for the survivors, and so a faster growth to maturity at six to seven years in both sexes.[9] This fact also suggests that maturity in this whale is a matter of size, not age; and that, during the heyday of the whale herds, even the huge bounty of the Antarctic zooplankton was significantly cropped down.

Baleen Whale Concentrations

The baleen whales depend on the wealth in food of the Arctic and especially of the Antarctic seas, to make a rapid recovery from breeding which they do during their annual migrations to tropical waters. Their feeding grounds can be deduced from the records of whale captures.[10] These are reproduced in Figures 13 and 14. It will be seen that, in the summer months (presumably both the northern and southern summers), the herds of baleen whales were hunted mostly in the Arctic Ocean north of Bering Strait, and also in the Norwegian Sea, Davis Strait and off Newfoundland. In the southern hemisphere they were hunted in a belt of ocean across the south Atlantic and Indian Oceans, up the east and west coasts of Australia, Africa and south America. In the winter, the chief whaling grounds were unmistakably in the Southern Ocean, all round the Antarctic Continent, where, as shown in Chapter 12, upwelling rich deep water nourishes a rich phytoplankton, which in turn nourishes many millions of tons of zooplankton, mainly the 'Krill.'

Sperm Whales

The sperm whales, toothed whales which hunt active prey such as fishes and squids, also live in the open oceans. They are able to make very deep dives, for there are many recorded cases where sperm whales

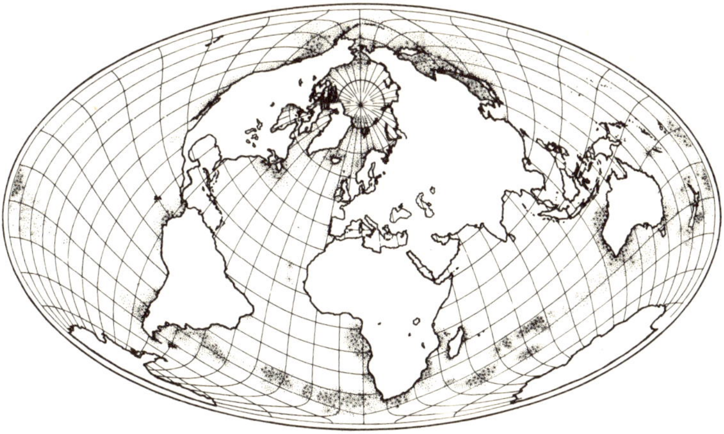

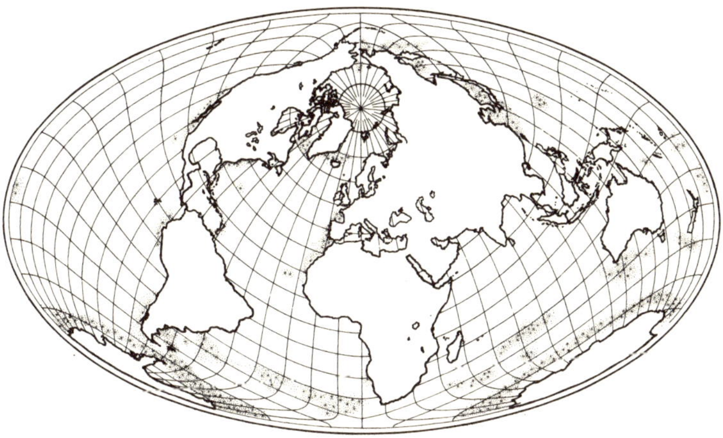

Figures 13 and 14 Baleen Whale grounds, summer and winter respectively. From Walford.[10]

have been found drowned by becoming entangled in submarine cables at depths as great as 1,000 metres. When they are about to make a deep dive, they sprint to gain speed, and after the dive may come up exhausted, lying panting at frequent intervals.[11] To help such dives, obviously a necessity in their hunt for food, sperm whales store fat of low density, for at great depths the air contained in the lungs becomes compressed and loses its buoyancy. The rate of heart beat also adjusts to economise air, and there are modifications to the lungs. They dive, when over the continental shelves, for bottom-living fish; but in the oceans they dive for squid, many of great size. At a depth of many hundred metres, light must be very faint, and the eye of the whale is not adapted for faint light; yet the whales successfully hunt and catch swift jet propelled swimmers such as the squids. They must use some form of sonar, as has been proved recently. Some squids are capable of speeds of 790 cm./sec. which is equivalent to about 15 knots; the top speed of the sperm whale is only 8 knots, though it can make short sprints of 10 to 12 knots.[12]

The fishery for the sperm whale has not yet had any serious effect on the stocks, as judged by the number taken annually. They are gregarious animals, each bull being accompanied by many cows. The fishery is worldwide and the same material can be used to show the distribution of the hunting. Sperm whaling took a hard knock when mineral oils took over the world's illumination for a while, before gas and electricity. Top quality candles were still made from products of the sperm whale, and a by product, ambergris, secreted in the gut of the whale to protect the intestine from the sharply pointed beaks of the squids it eats, is still very valuable in the cosmetics industry. Nowadays, too, the meat has high value. Lesser toothed whales such as the 'black fish' are also hunted.

Sperm Whale Concentrations

In Figures 15 and 16, maps show the position of capture of sperm whales in the summer and winter months respectively. At all times, whales were most abundant (as indicated by the catches), in the areas of upwelling of nutrient-rich water. Off southwest Africa and Peru, where the fisheries are so rich, the whales were also abundant. The squids feed on fish, and so no doubt do the sperm whales; and these will be most abundant where primary production is the greatest. The midocean belts of fertility due to the upwelling there of nutrient-rich water from a few hundred metres depth, are also regions of good hunting for the whales. The effect of the monsoons can be seen; for in the northeast monsoon in the Indian Ocean, the areas of upwelling off the Arabian coast and off the coasts of East Africa and eastern south

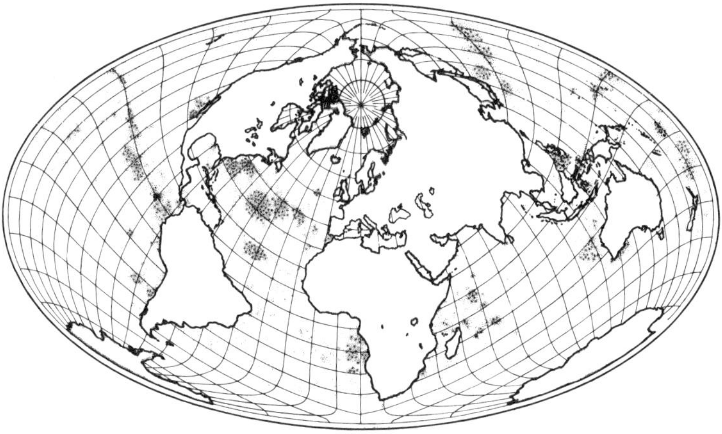

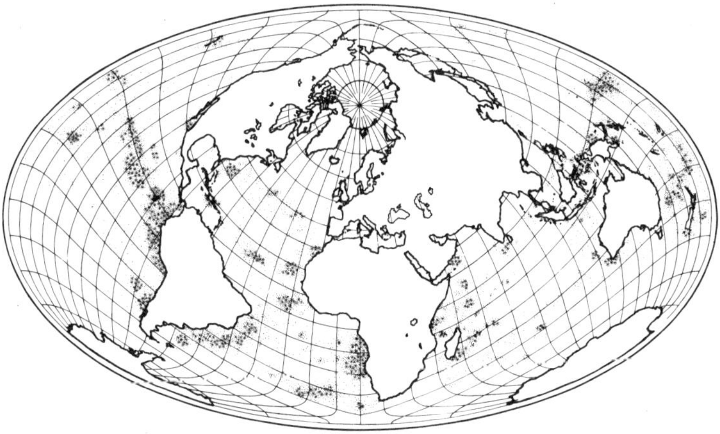

Figures 15 and 16 Sperm whale grounds, summer and winter respectively. From Walford.[10]

America are then areas of good sperm whale hunting. In the time of the southwest monsoon, between April and September, there was a big concentration of whales in the north central Atlantic, and also in the northwest Pacific; there were no good areas for whaling in the Indian Ocean during these months. While the rich areas off southwest Africa and off Peru are good hunting all year round, the sperm whales do not frequent the great areas of zooplankton fertility in the southern Ocean, as the baleen whales do.

The Floor of the Deep Ocean

Though the food supply in the ocean must grow scarcer with increasing depth and repeated use by successive layers of deep sea animals, everything must finally end on the ocean floor, whether detritus, dead animals and plants. In Chapter 11, the detritus was said to have little nutrient value as a food material for other animals, few of which are equipped to digest it; but that bacteria can easily digest detritus and convert it into good nutritious food material.

Bacteria are very abundant in the oceans, having there, as elsewhere, the ability to multiply rapidly up to the limit of the available food supply. Russian work has shown that the number, biomass and production of the bacterial flora of the bottom sediments diminish rapidly with the increase of depth,

Table 4 From Sorokin[13]

Depth metres	Nature of soil	Bacteria Present: Number x 10^6	Biomass mg.(wet wt.) l	Production mg. (wet wt.)l/day
25	Coral sand	2030	910	226
1100	Grey silt	105	51	3.3
5020	Diatom-Radiolarian ooze	37	17	1.4
5530	Red clay	7.1	3.4	0.03

No doubt dissolved organic matter is also present in the deep sea, because small plants have been found far below the reach of any daylight, living in healthy condition, and this could hardly happen unless they were living as saprophytes. The Pogonophora, or beard worms, which live in the mud of the deep oceans down to the deepest trenches, have no gut and live by absorbing dissolved organic matter, such as amino acids, and is has been said earlier that many molluscs can live in part in this way (p. 130).

The Danish 'Galathaea' expedition found an invertebrate biomass of 10 to 12 gm./m^2 at 7,000 metres in the Banda Sea among the Indonesian Islands, a figure as high as in some productive European waters. The chief reason was the great amount of debris, and especially of waterlogged wood and plant fragments derived from the surrounding islands.[2] They found, too, that beneath an area of rich organic production, the bottom fauna at many thousands of metres depth reflected this richness, no doubt by way of the downward food chain. Thus, the benthic fauna had a biomass of 0.5 gm./m^2 off southwest Africa, but 0.37 off the moderately rich ocean off the mouth of the River Congo, and only 0.2 off the poorly productive east African coast. Under the poorly productive areas beneath the central oceanic gyres, the mean biomass of the bottom fauna may be only 0.1 to 0.05 gm./m^2.

A Saprophytic Food Chain

Unlike the Russians, the Danes on the 'Galathaea' found a rich bacterial flora in the ocean depths, well able to break down material which still contained energy, such as woody material derived from the continents, and chitin, both resistant to normal digestion. The presence of numerous bacteria would give the bottom mud some value as food for the mud swallowers such as the sea cucumbers, molluscs and polychaete worms. The filter feeders would find material in the smaller drifting particles of detritus, and all these small animals would provide food for active crustacea, and they for fishes. At all events, the experienced Danes were surprised at the comparative abundance of life in the bottom of the oceans. They were also suprised at the amount of vegetable matter, logs, leaves, fruits, branches and bamboos found on the ocean floor at 4,000 metres and more. These decay slowly, nourishing a bacterial flora, and also a fauna of drilling and burrowing animals which can get energy from this predominantly carbohydrate material. Probably, as with our own wood borers, they get help in the digestion of the wood by an intestinal flora of bacteria. These boring and drilling organisms are food for other animals. A deep sea urchin was found to be grinding up the wood for the sake of the animals it contained.

This fauna of invertebrates on the floor of the deep oceans supports a limited fauna of predatory animals, chiefly fish, which must have the faculty of successfully hunting prey in total darkness. None of these fish are likely to be worth the interest of man as a predator. They are much too sparsely scattered to be harvestable. These lucky animals are therefore likely to be left alone to pursue in tranquillity their dark, cold and lonely way of life.

Oceanic Pelagic Fish

Many pelagic oceanic fishes are good food for mankind. The flying fish have been mentioned in Chapter 9; there are commercial fisheries for them in the West Indies and off the coasts of India. In both cases, the shoals of flying fish are attracted for easy capture by spreading on the sea surface a floating mat of palm leaves. A further attraction is the spreading of ground bait in the form of a loathsome bag of rotting fish from which a stream of stench and fragments drifts upwind as the boat lies to. Any passenger just holding his own against sea sickness in the wildy gyrating boat is likely to collapse when this 'fresh' is produced. Once the fish begin to collect under the raft, they can be caught by careful sweeps of a hand net. But they are easily scared off, and any approaching boat will be waved off with furious gestures. If the fish are few in number very fine cotton hand lines baited with tiny bits of fish will be used; as the lines are unweighted, it takes some skill to make a good cast against the trade wind. When a fish is hooked, it takes the only evasive action known to it; it spreads its wings and tries to take off. Often the palm leaves will be festooned with the bright red eggs of the flying fish, and sometimes these eggs, crushed in the hand, will be used as ground bait.

The dolphin fish, *Coryphaena*, has been mentioned in Chapter 13, and other edible members of the oceanic surface community include the jacks or crevallies, members of the Carangidae, or horse mackerel family; and large and strong members of the mackerel family such as the kingfish, the queenfish, *Cybium*, the rainbow runner, *Elegatis*, the wahoo; and the billfish, the sailfish and the swordfish. But the most important of the oceanic fishes are the tunnies or tunas. These fish have been appreciated in the Mediterranean countries from very early times, for they are easily preserved in brine, or by smoking or cooking, and are then a durable article of commerce. Their migratory habits have long been known; at suitable places, great traps of netting are set to intercept the migration and guide the fish into an enclosed pound from which they can be gaffed out. Similar traps are used in Japanese coastal waters.

Tuna Fisheries

There is an important fishery for bluefin tuna in Brittany, Spain and Portugal, by handlines trailed from spreaders. The Pacific islanders and those of the Maldive Islands have also long fished for the members of the tuna family and especially the bonito, attracting the fish with ground bait of small live fish, and then skittering a barbless hook in a feather jig among the milling tuna. This method has been modernised and mechanised by American and Japanese fishing outfits, large and fast ships, with refrigerated brine tanks and a tank on the stern able to keep alive the ground bait of small sardines or Atherines (silversides). The needs of the tuna clippers for live bait virtually fished out the anchovy *Cetengraulis mysticetus* in the Gulf of Nicoya, in Panama.

The American tuna fishery began at the turn of the century, but success did not come until the food shortages of the First World War. People became familiar with tuna meat, and promotion campaigns sent demand soaring. Purse seining with very big and heavy nets, aided by mechanical turntables and hoists spread from California south to central America and to Peru and Chile, thus covering the west coast of America and several hundred miles to sea. Meanwhile, the Japanese were developing their tuna fisheries in an easterly direction along the belts of the equatorial currents. Now, their tuna fisheries encircle the globe. They have developed the floating longline for the capture of oceanic tunas. These longlines may be many miles long, of heavy line to which, at intervals of twenty-five metres or more, baited hooks on long snoods are seized. The whole longline is suspended below the surface from buoys, the distance below the surface depending on the habits of the fish. The region of the thermocline, where the warmer surface layer changes suddenly to the colder deep water (Chapter 12) is a likely layer of sea to try. The tunas feed on oceanic macroplankton, fish, squids and crustacea, which feed on the locally-rich microplankton.

A Managed Tuna Fishery

The United States authorities became alarmed at the rate of exploitation of the yellowfin tuna, the largest and most valuable of the commercial tunas, and became interested in the management of the fishery, just as the halibut fishery was managed (Chapter 14). In 1940, the catch of yellowfin rose to 85,000 tons, and after the Pacific war, to 154,000. The catches of the smaller tunas such as the Alabacore and the skipjack showed that they were in no danger of overfishing, but the yellowfin was thought to be in danger. Management was eased by the knowledge, based on tagging experiments, that not one out of 6,000 tagged tunas was caught outside the areas of the American tuna fishery in the eastern Pacific. Based on the management of the halibut fishery, it was decided to limit the catch of yellowfin. In 1962 a catch quota of 83,000 tons was set for the year, but it was not possible to enforce this, since the fishery is multinational. The other countries, Mexico, Peru, Canada and the Central American States demanded a national quota. In 1968, the quota of yellowfin tuna was raised to 106,000 tons, which was soon fulfilled, and then the tuna fleets turned to the less endangered species, skipjack.[14]

Guyots

The oceanic environment also includes the guyots, or sea mounts, which are flat-topped submarine mountains. Little has yet been published about the fauna of these guyots, but their presence arising from the depths causes upwelling; echograms show large objects, perhaps large fish or squid, around and over many sea mounts between California and Hawaii.[15] On sea mounts in the eastern Pacific various fish, including a deep sea sole, rockfish and gurnard, have been caught; and on the 'Stranger Bank' in latitude 16° 52′N, longitude 130°47′W, 1,000 tons of yellowfin tuna and smaller quantities of skipjack tuna have been caught.

Corals and Coral Reefs

Coral islands, which do reach the surface, are well researched. There are three kinds of coral reefs; the fringing reef, which grows attached to the shore as a shelf; the barrier reef, which was probably a fringing reef but which has become separated from the land by a subsidence, leaving a more or less deep water channel between the reef and the land; and the atoll, one of the most striking features of the oceanic environment. An atoll is a circular reef, with one or many gaps which can be used for navigation, surrounding a lagoon which varies in depth from a few to sixty fathoms or more. Atolls are derived from fringing reefs which continued to grow upwards when the island on which they were established subsided.

The reef building corals are colonial animals which belong to the same group of animals as the anemones, and indeed each individual polyp much resembles a small sea anemone, with its circle of tentacles designed to stun and carry its prey, minute zooplankters chiefly, to the central mouth for digestion. But the coral polyps grow on branched stalks which secrete a sheath of solid limestone around themselves. On the reefs also grow calcareous algae, which are true plants, but which also secrete lime. These are the reef builders, which secrete the limestone of the reef and consolidate it. But there are many reef destroyers, such as all kinds of drilling and burrowing animals, which riddle and loosen the solid coral, and so allow the ocean swell to break off chunks which are either thrown onto the reef, thus raising it above the sea surface, or slide down the outer face of the reef to form a talus of debris. Ultimately, the reef may become raised high enough above the sea to collect soil and become a ring of low islands around the lagoon.

Among the enemies of coral reefs is the well-publicised Crown of Thorns starfish. This feeds, like other starfish, by everting its stomach over the mat of living coral polyps and digesting them, leaving areas of bare limestone. All starfish, even our own harmless ones, are liable to population explosions for no obvious reason; the Crown of Thorns was in any case much more common than had been suspected until the uproar called attention to it.[16] The plague occurred in widely separated parts of the ocean, but now seems to be dying down, as such populations do; in Guam, they are studying the regeneration of denuded reefs. Corals produce drifting larvae, called planulae, by the million, which soon settle on any suitable surface.

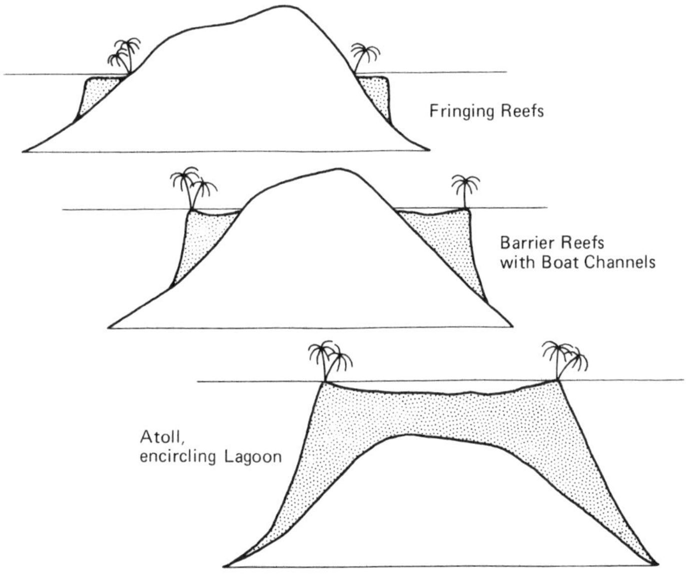

Figure 17 The three types of coral reef, showing Darwin's suggestion as to their formation, by the subsidence of the land.

Coral reefs can be harmed by man's pollution. In both Singapore and Eilat, on the Red Sea coast, reefs are being harmed. In Eilat, the fringing coral reef suffers from oil spills from an oil terminal, which smother the living coral. Phosphate ore is handled on a large scale at Eilat, and many algae are able to flourish in water enriched by crude rock phosphate dust, growing vigorously over the reef and smothering out the coral.[17]

Symbiotic Algae

There is much more plant than animal material present in a coral reef, indeed the ratio of plant to animal may be 3:1.[18] Not only are the reef building algae concerned; the coral polyps have a brownish or olive-green colour, due to small plant cells called zooxanthellae which are harboured in the tissues of the coral animals. Some of these zooxanthellae are stages in the life history of dinoflagellates, free living members of the nanoplankton. They live like plants, producing in sunlight new organic matter from carbon dioxide and other essentials provided by the host coral. The host uses these new organic materials for its own requirements. So dependent have many corals become on their productive guests that they will starve if denuded of the zooxanthellae, or kept in the dark where photosynthesis is impossible.[19] Plants and animal are in a state of mutually profitable partnership, and the coral polyps may even lose the ability to catch and feed on animal plankton. If a coral polyp is deprived of its zooxanthellae, it may secrete into the sea measurable quantities of phosphate, but will not do so when the plant cells are present and producing.

Experiments using labelled radioactive ^{14}Carbon, show that isolated zooxanthellae, when in vigorous growth, excrete about twenty per cent of the newly-fixed ^{14}Carbon to the medium in thirty minutes, and that this rate of excretion is doubled if the zooxanthellae are incubated in an extract of the host coral. The products of the zooxanthellae vary according to the materials supplied, so that it seems that coral polyps have some control over the photosynthesis of their plant guests.[20] All these experiments point to a relationship amounting to dependence between coral polyp and minute plant cells. Coral polyps may excrete much mucus, so much so that the water flowing over a reef may carry away the equivalent of two per cent of the total production of the reef community.[21] This mucus forms aggregates, just as does dissolved organic matter (p. 144), and can then become an important source of food for the filter feeding zooplankton near the reefs. Feeding trials with six species of scleractimian reef building corals from the Bismarck Archipelago, using materials labelled with ^{14}Carbon, have recently proved[22] that these corals are able to assimilate planktonic bacteria and dissolved organic matter from seawater. The coral polyps could daily assimilate organic carbon equivalent to 10-20 per cent of their own carbon content, from concentrations about the same as those found in nature. The giant clam *Tridacna*, common on coral reefs, also harbours captive algal cells, and is able to absorb dissolved organic substances from the seawater, just as can the less exotic common mussel, *Mytilus* (p. 130).

Primary Production by Coral Reefs

Because of their dense population of symbiotic plant cells, coral reefs can be highly productive communities. The high rate of primary production of the beds of turtle grass in the lagoons inside the Laccadive Islands atolls has been mentioned (p. 125); it was 3.38 $gm.C/m^2/day$. Yet this high rate was surpassed by that of the reef itself, which produced at 3.73 $gm.C/m^2/day$.[13]

But not all coral reefs are self-supporting. Fringing reefs, in particular, have rates of respiration higher than their rates of primary production; the balance must be made up from detritus and organic aggregates captured from the seawater by the tentacles of the coral polyps. The zooplankton, which would be the natural food for the coral polyps, can indeed be an important source of nourishment, but by itself cannot meet the needs of a coral reef community.

Finale

This book has followed fresh water from its origin as rainfall on the mountains, to its return home to the ocean, mother of all life, and the home of creeping things innumerable, both small and great beasts; the home, too, of that Leviathan, who takes his sport therein. A very little of this ocean is continuously distilled, to fly through the air and settle on the land. But even this little supports with a bounteous hand all life both in fresh waters, and on the land surfaces of the globe. The absence of water signifies death.

Strong and free, strong and free,
The floodgates are open, away to the sea.
Free and strong, free and strong,
Cleansing my streams as I hurry along,
To the golden sands, and the leaping bar,
And the taintless tide that awaits me afar,
As I lose myself in the infinite main. . . .
Undefiled for the undefiled. . . .

Charles Kingsley, *The Water Babies.*

NOTES

1. Hobson, L.A., Mengel, D.W. and Barber, R.T. 1973. 'Primary production and pools of organic carbon in the mixed layer of the sea.' *Mar. Biol.* 19: 298-306.

2. Braun, A.F. 1956. 'Animal life of the deep-sea bottom.' In *The 'Galathea' Deep-sea Expedition, 1950-2*. London, Geo. Allen and Unwin. 1956.

3. Wood, E.J. Ferguson. *The Living Ocean.* (In press.)

4. Merrett, N.R., Badcock, Julian and Herring, P.J. 1973. 'The Status of *Bentalbella infans* (Pisces, Myctophoidei), its development, general biology, and distribution in the Eastern North Atlantic.' *J. Zool. London.* 170: 1-48.
5. Blaxter, J.S., Wardle, C.S. and Roberts, B. 1973. 'Aspects of the blood and circulation physiology of deep-sea fish.' *Proc. 'Challenger' Soc.* IV (3): 140.
6. Hickling, C.F. 1963. 'On the small deep-sea shark *Etmopterus spinax* L., and its Cirripede parasite, *Anelasma squalicola* Lovén.' *J. Linn. Soc.* 45: 17-24.
7. Denton, E. 1963. 'Buoyancy mechanisms in sea creatures.' *Endeavour*, 22: 3-8.
8. Moffler, M.D. 1972. 'Plasmonics communication by radio waves.' *Hydrobiologica* 40 (1): 131-43.
9. Lockyer, Christine. 1973. 'Decline in the age of sexual maturity in southern Fin Whales.' *Proc. 'Challenger' Soc.* IV 9/0: 144-5.
10. Walford, L.A. 1958. *Living Resources of the Sea.* The Conservation Foundation. Ronald Press Co., New York.
11. Gaskin, D.E. 1964. 'Recent observations in New Zealand on some aspects of the behaviour of the Sperm Whale.' *Tuatara* 12 (2): 106-14.
12. Williamson, G.R. 1972. 'The true body shape of rorqual whales.' *J. Zool. London* 167: 277-86.
13. Sorokin, Y.I. 1971. 'Bacterial populations as components of oceanic systems.' *Mar. Biol.* 11 (2): 101-5.
14. Schaefer, Milner B. 1970. 'Management of the American Pacific tuna fishery.' In *A Century of Fisheries in North America.* Sp. Pubn. No. 7, American Fisheries Society, Washington D.C. 1970.
15. Hubbs, Carl L. 1959. 'Initial discoveries of fish faunas on seamounts and offshore banks in the eastern Pacific.' *Mar. Biol.* 6 (1): 2-17.
16. Weber, J.N. and Woodhead, P.M.J. 1970. 'Ecological studies on the coral predator *Acanthaster planci* in the south Pacific.' *Mar. Biol.* 6 (1): 2-17.
17. Fishelson, L. 1973. 'Ecology of coral reefs in the Gulf of Aqaba influenced by pollution.' *Oecologica* 12 (1): 55-67.
18. Moore, Hilary B. 1958. *Marine Ecology.* New York, John Wiley & Sons Inc.
19. Franzisket, Ludwig. 1970. 'The atrophy of hermatypic reef corals maintained in darkness, and their subsequent regeneration in light.' *Int. Rev. Ges. Hydrobiol.* 55 (3): 1-12.
20. Muscatine, L., Pool, R. and Cernichiari, E. 1972. 'Some factors influencing selective release of soluble organic material by zooxanthellae from reef corals.' *Mar. Biol.* 13 (4): 298-308.
21. Johannes, R.E. 1967. 'Ecology of organic aggregates in the vicinity of a coral reef.' *Limn. Ocean.* 12 (2): 189.
22. Sorokin, Y.I. 1973. 'On the feeding of some scleractinian reef-building corals with bacteria and dissolved organic matter.' *Limn. Ocean.* 18 (3): 380-5.
23. Qasim, S.Z., Bhattathiri, P.M.A. and Reddy, C.V.G. 1972. 'Primary production of an atoll in the Laccadives.' *Int. Rev. Ges. Hydrobiol.* 57 (2): 207-25.

INDEX

BIOLOGY AND ENVIRONMENT
General Editor:
Richard N. T-W-Fiennes

ECOLOGY AND EARTH HISTORY
Richard N. T-W-Fiennes

THE LIVING OCEAN: MARINE MICROBIOLOGY
E.J. Ferguson-Wood

TERRESTRIAL ENVIRONMENTS
J.L. Cloudsley-Thompson

MAN AND WILDLIFE
L. Harrison-Matthews, F.R.S.

MAN AND NATURAL RESOURCES: AN AGRICULTURAL PERSPECTIVE
Sir Cedric Stanton Hicks